An ecological framework for marine fishery investigations

FAO FISHERIES TECHNICAL PAPER

283

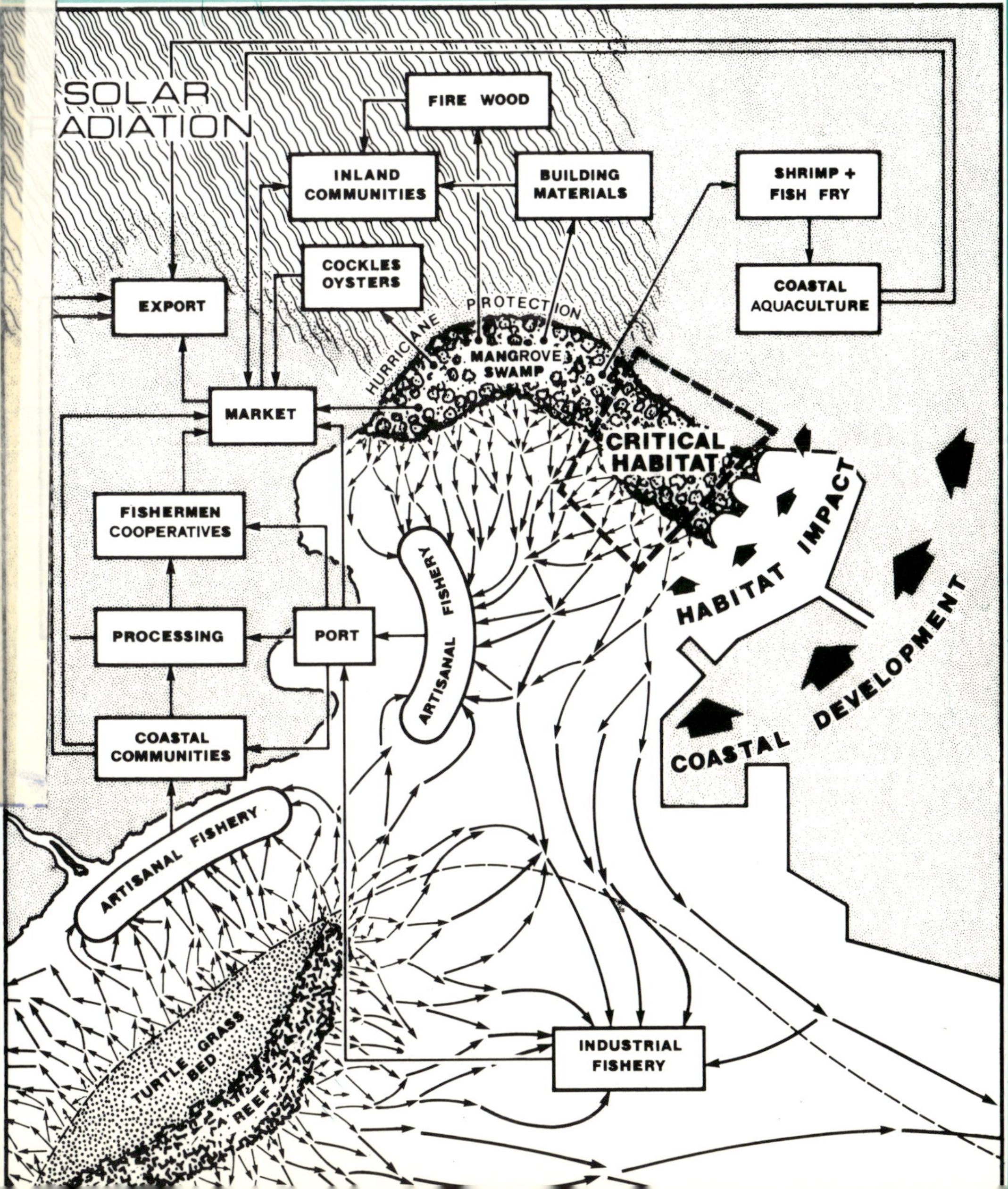

FOOD AND AGRICULTURE ORGANIZATION OF THE UNITED NATIONS

An ecological framework for marine fishery investigations

FAO FISHERIES TECHNICAL PAPER

283

by

J.F. Caddy
Senior Fishery Resources Officer
Marine Resources Service
FAO Fishery Resources and Environment Division

and

G.D. Sharp
Center for Climate-Ocean Resources Studies
PO Box 12294
Gainesville, Florida, USA

FOOD
AND
AGRICULTURE
ORGANIZATION
OF THE
UNITED NATIONS
Rome, 1986

The designations employed and the presentation of material in this publication do not imply the expression of any opinion whatsoever on the part of the Food and Agriculture Organization of the United Nations concerning the legal status of any country, territory, city or area or of its authorities, or concerning the delimitation of its frontiers or boundaries.

M-43
ISBN 92-5-102510-X

All rights reserved. No part of this publication may be reproduced, stored in a retrieval system, or transmitted in any form or by any means, electronic, mechanical, photocopying or otherwise, without the prior permission of the copyright owner. Applications for such permission, with a statement of the purpose and extent of the reproduction, should be addressed to the Director, Publications Division, Food and Agriculture Organization of the United Nations, Via delle Terme di Caracalla, 00100 Rome, Italy.

© FAO 1986

PREPARATION OF THIS DOCUMENT

First presented as background document of the ACMRR Working Party on the Assessment of Living Resources in near-shore Tropical Waters, Rome 28 February - 4 March 1983, some of the sections in this first document were extracted in a paper presented at the "Simposio Internacional sobre las Areas de Afloramiento mas importantes del Oeste Africano (Cabo Blanco y Benguela)" sponsored by Instituto de Investigaciones Pesqueras, Barcelona, 21-25 November 1983. Permission to extract sections from the conference paper is gratefully acknowledged. The original ACMRR paper, since revised and extended, is intended to provide access to the extensive literature on marine ecosystems, and in particular to some of the theoretical and practical implications of the concept of trophic interractions, which is becoming recognized as a vital prerequisite for understanding the dynamics of multi-species fisheries systems. As such, the present paper can be considered supplementary reading for those beginning work in various fields related to renewable marine resources, particularly stock assessment, whose calculations can hopefully be placed in a more realistic ecological context following study of some of the literature referred to in this document.

We have felt very conscious as authors of the difficulties of avoiding a subjective approach because of the "agglomerative" nature of much ecological research nowadays: separate schools of thought and practice with different ecosystems seems to be independently discovering and converging on similar principles - it is often just the time and space scales and objectives of the research that are different: it would therefore seem unwise to judge between them prematurely at this stage. As a result, the structure of the paper provides first (Part I) a very broad brush description of some of the main elements of fisheries ecology, which can be regarded as a self-contained account of the topic. The rest of the text (Part II) deals in rather more detail with individual topics touched on in the text, in a series of sections which can be read essentially independently of those preceeding and following.

ACKNOWLEDGEMENTS

The authors wish to express their gratitude to those many colleagues who have contributed to the work through discussions and exchange of ideas. Jim Kapetsky, Robin Welcomme, Jorge Csirke and Serge Garcia come particularly to mind; Ken Mann and Al Tyler also provided comments and encouragement, but cannot be held responsible for any facts or opinions expressed herein. Kathy Dorsey made an important contribution by reorganizing the first draft of this manuscript into a more coherent form. We would also express a debt of gratitude to the patient FAO typists who have suffered the many revisions to this document, especially Anna Rita Colagrossi, Emanuela Cheli, and Jackie Ellis, and to Gloria A. Soave who checked the bibliography. Sandro Cassola prepared most of the figures. The views expressed here are those of the authors, and do not necessarily reflect a formal policy on the part of FAO.

FAO gratefully acknowledges copyright permission to reprint the quotations on page v from:

THE LOG FROM THE SEA OF CORTEZ by John Steinbeck
Copyright John Steinbeck and Edward F. Ricketts, 1941
Copyright John Steinbeck, 1951
By permission of McIntosh and Otis, Inc.

THE LOG FROM THE SEA OF CORTEZ by John Steinbeck and Edward F. Ricketts
Copyright 1941 by John Steinbeck and Edward F. Ricketts
Copyright renewed (C) 1969 by John Steinbeck and Edward F. Ricketts, Jr
Copyright 1951 by John Steinbeck,
  renewed (C) 1979 by Elaine Steinbeck, Thom Steinbeck and John Steinbeck IV
Reprinted by permission of Viking Penguin Inc.

Distribution:

FAO Fisheries Department
FAO Regional Fishery Officers
Marine Sciences (general)
ACMRR Members
Authors

For bibliographic purposes this document should be cited as follows:

Caddy, J.F. and G.D. Sharp, An ecological framework for marine fishery investigations. FAO Fish.Tech. Pap., (283):152 p.
1986

ABSTRACT

Some of the key concepts of fisheries ecology are described in a broad brush interpretation, including recent developments in a variety of fields. The text is intended as supplementary reading for fisheries workers, especially in developing countries, who do not always have ready access to current literature on applied marine ecology.

An attempt is made to develop a wide range of concepts in a form that will hopefully encourage their incorporation into a practical, decision- making context. The food web and associated trophic interactions form the principal theme, in an approach that gives equal emphasis to qualitative, as well as the less easily measured quantitative considerations. An attempt is made to illustrate the consequences of the aggregated nature of much of marine production, as well as the subsequent dispersal of production in space and time, and how these processes affect the potential for economic harvest of commercial components of the ecosystem.

Separate sections touch on environmental influences on production, relevant spatial and temporal scales for ecosystem analysis, life history strategies, diversity and stability, the concepts of the ecological niche, the community and the assemblage, and outline some first steps towards quantifiying production in marine ecosystems. Different approaches to representing trophic and other interactions are discussed, with examples from the literature.

Reference is made to several ecological subsystems, in order to illustrate the main concepts presented. These include the mangrove ecosystem, the arcto-boreal macrophyte community, a mediterranean demersal fish assemblage, and the oceanic ecosystem associated with high seas tuna stocks.

In practical terms, it is concluded that the first and simplest approach to multispecies resource management is not necessarily the manipulation of individual food web components, but the identification, mapping and conservation of critical habitats, especially centres of local production, and their associated ecological dissipation structures.

"... Our own interest lay in relationships of animal to animal. If one observes in this relational sense, it seems apparent that species are only commas in a sentence, that each species is at once the point and the base of a pyramid, that all life is relational to the point where an Einsteinian relativity seems to emerge. And then not only the meaning but the feeling about species grows misty. One merges into another, groups melt into ecological groups until the time when what we know as life meets and enters what we think of as non-life: barnacle and rock, rock and earth, earth and tree, tree and rain and air. And the units nestle into the whole and are inseparable from it."

"... But all the fish actually were eaten; if any small parts were missed by the birds they were taken by the detritus-eaters, the worms and cucumbers. And what they missed was reduced by the bacteria. What was the fisherman's loss was a gain to another group. We tried to say that in the macrocosm nothing is wasted, the equation always balances. The elements which the fish elaborated into an individuated physical organism, a microcosm, go back again the undifferentiated macrocosm which is the great reservoir. There is not, nor can there be, any actual waste, but simply varying forms of energy. To each group, of course, there must be waste - the dead fish to man, the broken pieces to gulls, the bones to some and the scales to others - but to the whole, there is no waste. The great organism, Life, takes it all and uses it all. The large picture is always clear and the smaller can be clear - the picture of eater and eaten. And the large equilibrium of the life of a given animal is postulated on the presence of abundant larvae of just such forms as itself for food. Nothing is wasted; 'no star is lost'.

"And in a sense there is no over-production, since every living thing has its niche, a posteriori, and God, in a real, non-mystical sense, sees every sparrow fall and every cell utilized..."

John Steinbeck:
"The Log from the Sea of Cortez"

FOREWORD

It is important to recognize at the start of any discussion of the relevance of marine ecology to fisheries, that a degree of understanding of the ecological context within which a harvesting activity is taking place is essential if adverse impacts of these activities are to be minimized, and the systems ability to support productive human activity is not to be endangered. We should also be aware that exploited components of complex biotic systems are linked to other components, not all of which are harvested, but may be essential to the economic productivity of the system. These linkages need to be understood where at all possible since one fundamental ecological principle is that "you can't change just one thing": all ecosystem components are interlinked.

It is relevant also to note that although a "Maximum Sustainable Production" may be postulated for the system as a whole, this will itself vary in time, depending on the stability of the system, although this whole system variability will be less than that for each of its major components considered individually.

System stability itself is a function of a range of extrinsic factors acting through meteorological and hydrographic influences; each modulated in turn, by the geographic, especially bathymetric, configuration of the area of interest, and expressed through a set of characteristic species, which through evolution, have come to evolve their own mutual checks and balances.

In an ideal world, the first step towards gaining a "degree of understanding" of a fishery, would be to proceed from the general to the particular; starting with an overall description or "mapping" of the natural system, its geography and hydrography, and from a listing and distribution of the main taxa, their biology, migrations and interrelationships; and from a knowledge of the fisheries as well as the importance of the fishing ground for the whole spectrum of human activities, before focussing attention on individual components of key interest at this point in time. This is basically the approach we are advocating in this document: recognizing that although there are a variety of approaches to ecosystem analysis and management, starting with as wide a view as possible of the system will avoid, to the extent possible, major ommissions and a "tunnel vision" in developing hypotheses and approaches.

To a large extent, this is the way that fisheries science developed in "pioneering" areas such as the North Sea, where a long tradition of descriptive work preceded attempts to manage individual stocks; an activity that only got fully underway well after the 2nd world war. The large body of information on North Sea ecosystems accumulated during and since the last century was necessary to the pioneering syntheses of scientists such as Alistair Hardy in the 1920s-40s, whose description of the interactions between plankton and commercial fish stocks provided much of the early impetus for the development of the (then) new discipline of biological oceanography. In the last years of his life (he died in May 1985), there has been a revival of attempts to arrive at a quantitative synthesis of interractions between components of North Sea food web components that has dominated meetings of the International Council for the Exploration of the Seas (ICES) in recent years. To a significant extent, this document recognizes the correctness of Sir Alister Hardy's earlier conclusions on the importance of food webs.

For several important areas of the world's oceans this detailed early descriptive work involving mapping and data gathering on the resource and environment has never been carried out, and although the FAO Species Identification Sheets are now published for many tropical areas (Table 1), a knowledge of the biology of key components and their interactions is not yet available. Thus, the important industrial fisheries that sprang into existence, often as late as the 1960s and 1970s as a result of technology transfer from high latitude fisheries, usually lacked a proper basis of biological information for their control and management, often with serious consequences. The unit stock and single species assessment rationales first used in attempting to manage these fisheries, notably the generalized production model, and more recently, short-cut methodologies based on broad generalizations on growth and mortality, have mostly considered the exploited species in isolation. This has obvious drawbacks in complex tropical systems, and has accentuated an interest in multispecies fisheries theory. The latter is still, however, in an early stage of development, although recently attempts have been made to address this question (e.g., Pauly and Murphy, 1982; May, 1984). Good work is now being done, but the search for a suitable theoretical framework is largely inconclusive to date, and in many tropical areas at least, seems to have pushed ahead of the necessary basic field studies and experimentation that preceeded similar work in the simpler high latitude ecosystems.

The need for a basic understanding of fisheries ecosystems and their interactions, especially in the tropics, is an urgent one, and recent conferences have been useful especially in examining the physical systems themselves within which important tropical fisheries take place.

It is hoped that even though our priorities and interpretation of the field may not meet with universal agreement, that the present compilation of concepts and references in this field will be useful in providing workers unfamiliar with the literature on marine ecosystems, with some access to this extensive body of information, that may help to place their fisheries problems in a proper ecological context.

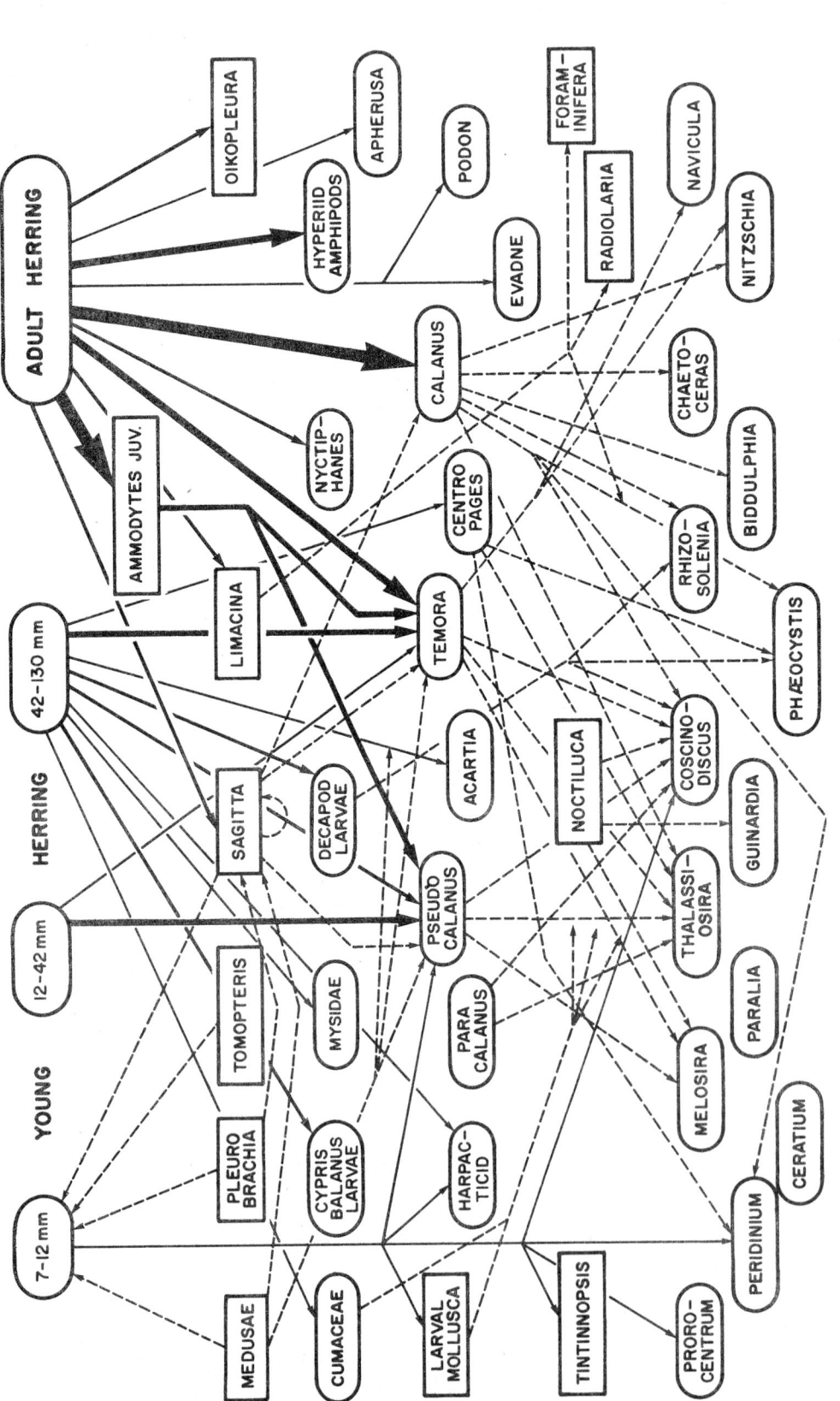

Figure 1 Hardy's herring feeding diagram. An early diagrammatic representation of the feeding relationships in the North Sea plankton. As used by Hardy (1924), the arrows point from the predator to the prey, implying a controlling influence of predators on prey biomass. This may not always be true, and recently the trend has been toward showing the flow of materials from the planktonic organisms to the herring biomass, and vice versa for some pelagic predators on larval herring: (thickness of arrows being roughly related to the degree of influence on the prey, or (in the opposite sense), to the proportion formed by each prey in the diet of the predator, herring)

CONTENTS

	Page
INTRODUCTION	1

PART I: INTRODUCTION TO THE MAIN CONCEPTS DISCUSSED, AND TO SOME KEY IDEAS IN FISHERY ECOLOGY — 2

1. THE DIVERSE ORIGINS OF ECOSYSTEMS CONCEPTS — 2
2. BIOTIC AND ABIOTIC FACTORS IN MARINE ECOLOGY — 5
3. MOVEMENTS OF ENERGY IN SPACE AND TIME: TROPHIC LEVELS, OR FOOD WEDS AND PYRAMIDS OF NUMBERS? — 5
4. CHANGES IN TROPHIC LEVEL IN THE LIFE HISTORY — 10
5. "BOTTOM UP" OR "TOP DOWN" APPROACHES TO ECOSYSTEM ANALYSIS? — 13
6. QUESTIONS OF BIOMASS AND SIZE OF ORGANISM — 13
7. SOME PROPERTIES OF FOOD WEBS — 15
8. ZOOGEOGRAPHY AND LIFE HISTORY STRATEGIES — 19
9. LIMITATIONS ON FOOD WEBS IMPOSED BY AVAILABLE SURFACES — 19
10. FISHERIES ECOLOGY AND ASSESSMENT OF RESOURCES — 20

PART II: ELABORATION AND DEVELOPMENT OF SOME KEY CONCEPTS RELEVANT TO ECOSYSTEM MANAGEMENT — 24

1. TEMPORAL CHANGES AND STABILITY OF FISHERIES SYSTEMS — 24
2. SYSTEM STABILITY AND MANAGEMENT — 28
3. SPATIAL CONSIDERATIONS: MAPPING FISHERIES RESOURCES — 30
4. FISH PRODUCTION PER UNIT AREA ESTIMATES — 35
5. COLLECTING AND ANALYSING DATA ON FEEDING PREFERENCES OF COMMERCIAL FISH SPECIES — 36
6. EQUILIBRIUM CONCEPTS AND THE FLOW OF ENERGY THROUGH A SYSTEM AND ITS OPTIMAL UTILIZATION — 39
7. QUALITATIVE CONSIDERATIONS IN ECOLOGY – THE MANGROVE ECOSYSTEM AND CORAL REEFS — 43
8. DIAGRAMMATIC REPRESENTATION OF LINKAGES WITH SPECIAL REFERENCE TO FOOD WEBS — 51
9. QUANTIFYING PRODUCTION WITHIN THE FOOD WEB: SOME PRELIMINARY APPROACHES — 57
10. SIMPLE CATEGORIZATION OF LIFE HISTORY STRATEGIES: r- AND K- SELECTION AND THE ECOLOGICAL NICHE — 72
11. FOOD WEB ANALYSIS IN PRODUCTIVE COASTAL ENVIRONMENTS: THE COASTAL KELP, SEA URCHIN AND LOBSTER SYSTEM IN HIGH LATITUDES, AND SEAGRASS NURSERY AREAS IN THE TROPICS — 83
12. FOOD WEBS IN LOW AND HIGH ENERGY SYSTEM: A MEDITERRANEAN DEMERSAL FISH COMMUNITY VERSUS FISHERIES OF UPWELLING SYSTEMS — 86
13. INTERACTIONS IN A PELAGIC ECOSYSTEM: INFERENCES FROM FISH PHYSIOLOGY AND BEHAVIOUR STUDIES — 94
14. CONSIDERATIONS OF ECOLOGICAL EFFICIENCY AND LIFE HISTORY STRATEGIES — 102
15. DIVERSITY AND STABILITY OF FISHERY ECOSYSTEMS: ARTEFACTS OF SCALE? — 107

		Page
16.	SAMPLING THE MARINE ECOSYSTEM: THE CONCEPTS OF CONTAGION, COMMUNITY AND ASSEMBLAGE	113
17.	THE EFFECTS OF TEMPERATURE AND BODY SIZE ON FEEDING AND ON THE NATURAL MORTALITY OF PREY	119
18.	ESTIMATES OF NATURAL MORTALITY RATES FROM WHOLE SYSTEM VARIABLES	124
19.	SOME IDEAS FOR REPRESENTING FOOD WEBS IN THE FISHERIES CONTEXT	127
20.	SINGLE SPECIES MANAGEMENT IN AN ECOSYSTEM CONTEXT	127
21.	IMPACTS AT THE ECOSYSTEM LEVEL	130
22.	REFERENCES	135

INTRODUCTION

Marine biologists entering the field of fish population dynamics and fish stock management for the first time, are often struck by the small fraction of our knowledge of marine biology currently being used in assessing marine fish stocks. There are historical reasons for this to be touched on later, but it is clear that fish stock assessment as a practical discipline, rests on the assumption that quantitative values can be given to the extent and rate of fish removals from an ecosystem. Therefore, to a large extent, the tendency is to collect and analyse data on the quantity, size and age structure only for those fish stocks currently under active exploitation. The limitations of this approach are evident enough to field biologists, even for the relatively simple ecosystems of north temperate regions where much of fish population dynamics theory evolved, and some principal research concerns in this area are currently with environment-species, and interspecific interactions. These latter questions have of course been preoccupations of biologists for a long time, but although methods of quantifying food and feeding of fish have been well documented, e.g., Bagenal (1978), ways of representing and analysing the mass of data, especially on stomach contents, have been slower to develop.

The strength of the historical approach of arguing always from easily quantifiable variables, cannot be challenged, and considering that fisheries managers are by and large non-scientists, who are impatient with either qualitative or probabilistic statements, the methodologies that have tended to be adopted in fisheries are those that provide quantitative (deterministic) conclusions. An acquaintance with marine systems however, shows the value of understanding species interrelationships, and indicates that usually the best one should hope for is a statement giving rather wide probabilities around any quantitative result. Evidence is mounting that attempting to manage individual species independently is rarely feasible in the long term, especially for complex and diversified tropical systems, and we already know that the physical environment is variable in most ecosystems. With this realization, it is necessary to look for whole-system approaches which, despite having a high degree of uncertainty, also meet the criterion that they provide some qualitative or semi-quantitative guidelines for managers.

Unfortunately, as noted in the preamble to this manuscript, there are no simple, unambiguous guidelines or a single simple theoretical basis for management of multispecies fisheries (see e.g. Pauly and Murphy, 1982; and May 1984, for a sampling of the current range of opinions), and a high priority in practice must be given to empirical, descriptive approaches (e.g., Rigler, 1982). It is unfortunate also that the present state of development of marine ecological theory today does not lend itself to a simple narrative, since the discipline is by nature complex and rapidly developing, with a variety of modelling frameworks now in process of evaluation (see e.g., reports of the Dahlem Conferenzen, (Gulland and Garcia, 1984), the ICLARM/CSIRO Workshop (Jones, 1982; Platt, Mann and Ulanowicz, 1981). Nonetheless, we attempt here to put the more important ecological concepts in an overall framework, with the guiding principle given in Part II that the flow of organic matter and energy through oceanic food webs oceans is governed by basic thermodynamic principles. This does not imply however (Part II (1 and 2)) that we are dealing with thermodynamic stability - quite the contrary! Synthesized organics follow complex pathways (the food web), which are inevitably affected favourably or unfavourably, to a greater or lesser extent, when man harvests one or more components of the system.

Predicting the nature, extent and direction of such disruptions, is by and large, a matter of educated guesswork at present, although some clues are now emerging. As a general statement however, we feel sure that any future improvement in the way marine resources are managed will only be possible if more ecological concepts are incorporated into fisheries management than at present, and that such considerations should be especially taken into account when contemplating significant changes in the way that a fishery is pursued.

We believe that familiarity with some of the main concepts of marine ecology is essential background for all those involved in management of marine resources, and even more urgently, for those concerned with stock assessment in developing countries, to whom this literature on applied ecology is often unavailable.

The main rationale then of this document is three-fold:

1. To introduce fisheries workers, especially in developing countries, to a number of important schools of thought in marine ecology and their more recent extensions.

2. To point the way to some possible alternative approaches which may extend existing ecological theory to the aid of fish stock assessment, especially in tropical and semi-tropical areas where multi-species approaches are essential. A more "ecological" or holistic approach to data collection seems necessary than appeared earlier to be the case.

3. To place particular emphasis on the concept of the food web, as an aid to scientific management: how to construct food webs and what tentative conclusions may be drawn from them.

Great emphasis has been placed throughout on a visual approach in the belief that even complex interrelationships can be more readily grasped in this way by a reader unfamiliar with mathematical notation. The "flow diagram" in particular, in its many practical applications in fisheries, proves a key to integrating many different disciplines in a way that is not readily possible from a linear textual account. The other easily visualized, but often-neglected, spatial relationships in fisheries are treated in Part II (3).

The text is therefore intended both as background reading for courses in fish stock assessment, and as background material to aid in decisions and priorities in fisheries research and management science. At the same time, the preliminary nature of many conclusions in marine ecology is stressed, in that a definitive framework for action that applies in all situations does not yet exist.

PART I - INTRODUCTION TO THE MAIN CONCEPTS DISCUSSED, AND TO SOME KEY IDEAS IN FISHERY ECOLOGY

1. THE DIVERSE ORIGINS OF ECOSYSTEMS CONCEPTS

Because of the great diversity of ecological systems in nature, and the limited interchanges of information between ecologists studying the various terrestial, freshwater and marine ecosystems, a wide spectrum of theoretical approaches has developed over the last few decades, each with its particular methodology, terminology, theoretical structure and literature. Various concepts derived from other biological disciplines and ecosystems, and from studies of energy transfer and species interactions, have been gradually entering the marine fisheries literature. Many of these have subsequently undergone practical and theoretical development, and have become the bases for extant methodologies.

Some examples are:

(a) Statistical theory developed principally in agricultural research (e.g., Fisher, 1960) gave some of the early impetus to design of quantitative sampling procedures in fisheries and other applied biological sciences.

(b) Measures of dispersion and contagion have been especially well developed in plant ecology (e.g., Greig-Smith, 1964), and have also been adopted widely in benthic ecology (See Pielou, 1969). There have been some applications in fisheries, however most been concerned with sampling problems (Hewitt, 1981; 1982; Sharp, 1980a) and benthos (e.g., Saila and Goucher, 1966), and Part II (16).

(c) Benthic ecology itself has seen an early development of the concept of marine and freshwater communities (Thorson, 1957), which has been extended more recently to marine fish communities (e.g., Day and Pearcy, 1968; Tyler, 1971, Smith and Tyler, 1975). In general, however, other than for more obvious and discrete ecosystems (such as mangrove swamps, coral reefs) where the boundaries of the community are easily recognized, the concept of a "community" of organisms seems vague and tautological, and is tending to be replaced by two alternative concepts: that of the food web comprising trophic interactions, PART II (8, 9, 11, 12) and that of the species assemblage, expressing regularity of joint occurrence of species in the face of common environmental influences (PART II (16)).

(d) The continuously evolving concept of the ecological niche (Fry, 1971; Hutchinson, 1965; Kerr, 1974; 1980) has been influential, particularly in the study of complex tropical fish communities (e.g., coral reef communities), where the definition of the "environment" of each species is not simply given in terms of its physical surroundings, but also in terms of the physiological requirements of organisms (PART II (10)), although a simpler more mechanical definition of the niche may now be coming back into vogue for some purposes (Caddy, in press).

(e) Disciplines introduced first in the field of marine ecology, and limnology and hydrography, such as population energetics (Sharp and Francis, 1976), community diversity and ecosystem stability (e.g., Margalef, 1968; Welcomme, 1979), and the idea of the succession of communities toward a climax (Odum, 1969) are beginning to enter the fisheries literature with growing frequency. There is still however, considerable uncertainty about the proper interpretation of information on diversity and stability for two main reasons: the first, because we are aware that the many measures of diversity are dependent on the efficiency (and frequency) of sampling of a community: this makes such measures less than absolute or even replicable as numerical values. The second, because we are aware that diversity (and stability) are secondary or even tertiary parameters, whose values are reflections of more direct and simply measurable parameters, especially population density and/or environmental gradients and boundaries whose fluctuations in time and space are not readily perceived in the marine environment, but give rise to a sequence of diversity values as ecological succession takes place. Some alternative concepts that seem relevant here are presented in PART II (15 and 16).

(f) The concept of Q_{10} reflecting the rate of change of activity of organisms in response to temperature change, and the "envelope" of physiological parameters that define an organisms niche or range, has been widely developed by fish physiologists, and has entered the ecological and fisheries literature. A short introduction to the impact of this, one of the more important variables having an effect on fisheries systems, is given in PART II (17).

(g) Perhaps evolved initially as an extension of work carried out in medicine, animal husbandry, and various branches of physiology, the application of thermodynamics to the analysis of the food and energetic requirements of organisms has become a major preoccupation of both modern industrial societies and, in parallel, of marine scientists. In this context it can be used to throw light on the relative importance of different interchanges within the food web: (See PART II (6 and 9)).

(h) The need to measure, survey and monitor the impact of pollutants, nutrient enrichment, modification of water flow and other stresses of modern society on living resources, has led to the invention of many sets of indices which together are helpful in to discriminating among the effects of various stresses on marine organisms. With more understanding of a particular system, more complex indices that relate stress to biological responses can be developed (e.g., Schlesinger and Regier, 1982).

(i) Two alternative paradigms have been vying for the attention of marine ecologists in the past decade, namely, the classification of ecosystem by their trophic level (Lindemann, 1942), or by their size spectrum; (the pyramid of numbers of organisms with size present in the ecosystem). The latter conceptual framework, which follows from the early work of Elton on terrestrial ecosystems, and from such later workers as Platt and Denman (1975), seems now to be eclipsing the former.

Eight main themes with varying degrees of interrelationship seem to be of growing relevance to fisheries resource scientists:

1. The concept of the ecological community, i.e., those organisms from micro to macro scale that commonly occur together (not always because of trophic inter-relationships).

2. The concepts of trophic interaction, community energetics and their whole system representation in the form of the food web; i.e., the dynamics of those groups of species that are linked as predator and prey. The development of this theory in terms of fixed trophic levels, though possibly applicable to some terrestrial ecosystems, will as we shall see later, need serious modification in marine ecosystems.

3. Related to the study of trophic interactions and food webs, is the realization that in functional terms, most species fall somewhere in the spectrum between complete generalists and specialists in their feeding habits. Marine organisms also fall somewhere in the spectrum between small, rapidly growing and short-lived species, and larger, longer-lived ones with numerous overlapping generations (respectively referred to as r and K strategists; MacArthur and Wilson, 1967). Many aspects of biology and ecology can be related to these categories, which although qualitative, provide a useful but not self-sufficient, conceptual framework for considering life history strategies. A brief review of this field is given in Part II (10). In fact, a consideration of evolutionary strategies leads toward the fields of species diversity and population genetics - fields which cannot be dealt with at length in the present review, but which are relevant for the design of sampling strategies, and for the definition of unit self-reproducing populations and stocks: the main units for population dynamics and stock assessment calculations.

4. Because certain sampling gears tend to collect specific groups of species more often than would be expected from random sampling, these groups have been labelled as "species assemblages". These groups of species tend to be of the same size and sometimes of similar trophic level, but are not necessarily closely interrelated trophically, as in the food web concept. The trend is to regard their co-occurrence as stemming largely from common responses to gradients within abiotic environmental parameters (Knight and Tyler, 1973), such as depth, temperature and/or bottom sediment types, rather than from any mutual "attraction" or functional interrelationships, as is implied in the community concept. Some of these concepts and approaches are, as mentioned earlier, touched on in PART II (16).

5. The concept of the ecological niche has evolved from the original idea of a spatial property (the "place" where a species is found), to a much broader concept in which the physiological requirements of the species at different stages in its life history are taken into account. More recently, the niche concept has also come to include less easily defined biotic factors as well as behavioral considerations, e.g., competition, species interactions, etc.; in the

process suffering an inevitable dilution, so that the definition of a niche now potentially includes the infinite number of variables that could be measured in any particular environment (Kerr and Ryder, 1977) and in fact becomes indistinguishable from the word "environment" in its broadest sense. The case for also retaining a more restricted physical definition of the niche is touched upon in Part II (10).

6. There is often a clear contrast in evolutionary strategies between species that have evolved to withstand a sedentary or territorially limited existence, involving low energy expenditure in areas where suitable food organisms are available for much of the year, (examples: oysters, reef fish), and species which through migration involving high energy expenditure, are able to maintain themselves in a series of transitory niches, where conditions suitable for feeding and reproduction are temporarily available. This may lead us to classify the commercial or potentially commercial species into transients and residents: the first of these, although only present at one time of the year, may play an important role in relation to movement of biomass and food energy from local areas of high production (where trophic interaction with resident species occurs) to or across other areas of low inherent production (Sharp, 1980b: Part II (3).

7. The realization, following Ilya Prigogine and his co-workers, that the existence of ecological complexity is a direct function of the lack of equilibrium in natural systems (Nicolis and Prigogine, 1977; Prigogine, 1978), has its obvious implications for management of resource harvests from these systems, and this perception is certainly likely to be a major influence on ecological thought over the next few years at least. Gallucci (1973) summarizes the anomaly by noting that in an evolutionary sense, organisms and communities tend toward greater organization and structure... However, the second law "(of thermodynamics)" requires isolated macroscopic systems to evolve in time to a final state of maximum entropy "(disorder)". In fact, as Prigogine notes, there is a kind of inertial property of non-equilibrium: when reached, the system settles down to the state of "least dissipation of energy", leading to the formation of "dissipation structures": prime examples of these being food webs, which result in the chemical energy of photosynthesis being slowly released in an orderly fashion in time and space.

8. In this connection, the "equilibrium yield" concept in fisheries assessment is an important tool for allowing simple calculations to be carried out which allow a first estimate of the impact of changes in fishing strategy, assuming a relatively stable flow of energy through the system. However, it follows from the concept of dissipation structures that we are dealing with metastable states, dependent on a more or less constant flow of nutrients to a harvested section of the food web, and in this manuscript we point to certain important limitations to the use of this concept as an overall panacea for simplifying fisheries yield calculations.

The relevance of the above conceptual 'fields' to management of multispecies fisheries lies in the constant changes of the catch compositions that occur geographically seasonally, and over the long term. As each species has specific requirements and interactions within an ecosystem, the numerous adjustments in the ecological balance and displacements which we perceive only as changes in catch, must be taken into account, since it is very unlikely that fishery science will ever achieve even moderate success in understanding and rational utilization of aquatic resources from catch statistics alone.

Spencer Apollonio, Commissioner, Department of Marine Resources, State House, Augusta, Maine, developed this theme in the following way with respect to New England Fisheries:

"<u>A better understanding of the way the system works is clearly needed. The recent emphasis on quantitative assessments as the basic management tool has not been very helpful to managers</u> because assessments contain little predictive content... Perhaps a more qualitative understanding - which I believe is possible - of the characteristic functions of the system would be more useful for managers and less expensive. A manager frequently is not primarily concerned with an accurate estimate of the abundance of a stock. <u>He is more likely concerned with a probable trend of relative abundance.</u>"

He continues:

"Even grossly qualitative forecasts may be most useful management tools and serve to re-establish belief in the management process. For example, the probably valid prediction that increased fishing effort will lead to an increasingly variable stock abundance can be tested by the industry itself and thus enhance the credibility and thus the usefulness of management. Similarly, by taking full cognisance of the biological characteristics of the species, it should be possible to state which species are legitimate candidates for efforts at stock stabilization, and which are inherently highly variable. The industry believes that the latter is true of many species and it is probably right on at least some of them. Current management practice does not seem to acknowledge the probability - but instead seems to promise sustained yield for all species, regardless of their biological peculiarities, if only the proper regulations are promulgated. It is almost as though managers consider all species biologically equivalent and that the species characteristics are without management significance. ..."

"...If a manager, concerned with stocks, people, economics, and regulation, is to meet his responsibilities, he must have advice on how the system works. Species cannot be managed in isolation, just as many of them are not caught in isolation. The manager must understand interspecific relations if he is to manage interrelated fisheries. He must understand the nature of normal variations in abundance if he is to judge which species can be managed and at what cost. He must understand how the system operates if there is to be any hope of forecasting stock abundance. These are the capabilities we hope to attain through fisheries ecology."

A brief review of the main subject areas covered in the paper is provided in the rest of this section, with what the authors consider the more promising directions to follow in investigations of marine resource ecology amplified in Part II. Reference is made to the literature should a more concrete description or illustration of the concepts involved be necessary which is beyond the scope of the present text.

2. BIOTIC AND ABIOTIC FACTORS IN MARINE ECOLOGY

The impact of the environment on marine organisms is a major preoccupation in the marine ecology literature, and such impacts have been classified into abiotic: the impact of physicochemical factors: (temperature, salinity, depth, etc.), and biotic: (the impact of other organisms on the species in question). Part II (17) reviews some relationships in the first category. Biotic effects may be classified into trophic interactions (see Part II (5, 8 and 9)) in which predation is the major factor; and non-trophic interactions (Part II (16)). In this latter category fall other forms of interaction, such as competition for space, or even on occasions, mutual assistance (e.g., parasitism, commensalism, symbiosis); although for a discussion of the details of ecosystem interrelations, a text on marine ecology (e.g., Kinne, 1984) should be referred to.

In freshwater systems (e.g., food chains of essentially similar lakes), replication of observations allows the impact of environment to be quantified, and its variance to be estimated in a way that is rarely possible in the marine system. Thus, we are far here from the application of so-called "morphoedaphic indices of fish yield" developed by Regier and Henderson (1973), Schlesinger and Regier (1982), which allow production to be measured as a function of depth, climate, dissolved solids etc., without a great deal of other biological data. Marine systems have to be classified from biological and fisheries data, although comparative studies are now producing useful generalizations.

In addition to the biotic influence a species exerts on others, we can recognize that some organisms can have a marked impact on the environment or milieu they live in: here we include corals, mangroves, etc., which may be referred to as "substrate" species, and are often the descriptive term used to characterize a given community. Examples of community interactions of this kind are given in Part II (7) and (11): the coastal mangrove system in the tropics and subtropics, and the similar but less complex kelp-sea urchin system of higher latitudes. By contrast, a pelagic food web, the tuna-dominated system characteristic of world oceans, is touched on in Part II (13); and in Part II (12) a moderately complex demersal fish community from a low-energy environment is contrasted briefly with simpler food webs from high energy upwelling systems. In Part II (8) we illustrate various graphical approaches to modelling ecosystems, and note that such simple visual representations of reality play an important role in understanding systems with many interacting components.

3. MOVEMENTS OF ENERGY IN SPACE AND TIME: TROPHIC LEVELS, OR FOOD WEBS AND PYRAMIDS OF NUMBERS?

The idea of the trophic pyramid and the food chain were developed in terrestrial ecology, where food linkages are generally more specific and simpler than in the marine environment, but the same general approaches and vocabulary have now been widely applied in marine biology, and an extensive literature now exists in trophic studies in the environment. (See References).

The need to understand marine systems in their own right and not just apply terrestrial models is evident (e.g., Carlenton, 1985), and over the last half century or so, the "information gathering" phase that preceeds modelling, has began to show the real complexity of marine ecosystems.

In the earlier theory of food chains, organisms were classified into discrete levels according to the number of steps in the food chain they are above the primary producer level. This idea of a 'food chain' with a discrete number of steps was later modified to that of a 'food web' with each species being assigned a trophic level which was not necessarily a whole number. This approach too is coming to be regarded as an over-simplification for three main reasons, namely (a) there is a parallel detritus web that begins again from bacterial biomass supported by organic detritus coming from all components of food webs. It is difficult, or impossible in practice, to assign one trophic level to all individual species; (b) the trophic level of a species changes with age and season for many marine organisms; (c) as noted by Cousins (1985), from the viewpoint of scientific procedure the concept of a numerical trophic level is fatally flawed, since as a variable, it can only be measured for a species by restating the question as to the trophic level of its prey species, and so on down to the primary producer level. In other words, for the question to be answered, it requires a full knowledge of the trophic level of each 'subordinate' species. Even if this can be achieved, it is not obvious that this information could be extrapolated to the same species elsewhere, or to similar species in the same locality. This problem is of course more serious 'higher' in the food web, where 'higher' (in the food web) is here used in the relative rather than absolute sense (which seems quite sensible); but also occurs lower down. (See Figure 1 for an early example of the complexity of the problem.) Thus the Peruvian anchoveta is both a herbivore and primary carnivore feeding on zooplankton, as well as feeding on its own eggs while they are in the plankton; (which could make it, in one sense, a secondary predator?). What is its trophic level? Such questions abound, and even if answerable, have limited value. It is more relevant to draw attention to the fact that all three food organisms occupy roughly the same size range. This latter observation was the basis for the earlier approach of Elton, who expressed ecosystems in terms of the "pyramid of numbers". This concept in somewhat more refined form, is now becoming a focal point for research in biological oceanography, and points to the fact that smaller creatures grow and reproduce faster than larger ones, and it is this size-related aspect of life cycles that determines that there are enough small animals for large animals to eat, and this fact 'drives' the whole marine ecosystem, and finds its expression in the sort of relationships shown in Figure 2(A).

Figure 2   Illustrating the apparent similarity but fundamental conceptual difference between (A) Elton's "pyramid of numbers" (with size), and (B) the concept of trophic level (see text)

We have already referred to one pragmatic way of classifying those groups of species that occur together in an area more regularly than expected by chance alone, namely by <u>species assemblage</u>. We may note that although some members of an assemblage are undoubtedly linked trophically, all species in an assemblage are not necessarily so, even though they may show other interactions: they are primarily defined statistically by their degree of co-occurrence. Considering only trophic interactions, we generally refer to that group of species in an ecosystem that are linked with each other as a <u>food web</u>, which will ultimately include every species in a marine community. Marine food webs usually differ somewhat from simple terrestrial food webs, in that a given species will often occupy different points in a food web at different stages in its life history: the number of cross linkages are also much more diverse, and will change from location to location, and time to time.

Figures 3 and 4 illustrate in a very schematic form, the way in which production of organic material tends to be spatially concentrated in fairly restricted areas, both inshore; (nutrient material and growth factors entering the marine system from rivers, and generated by salt marshes, mangrove swamps and coastal seaweed or eel grass beds), and in the marine milieu proper, by phytoplankton production. The latter component tends to be more intense in areas of upwelling where production-limiting nutrient salts are replaced from lower down in the water column; and at the boundary between oceanic water masses where similar effects occur. Dispersal, predation and active migration result in organic material and living organisms being spread horizontally and vertically from these centres of intense production. These centres of production can be regarded as forming the centres of "dissipation structures", where biological material is synthesized and slowly dispersed outward. (Figure 3).

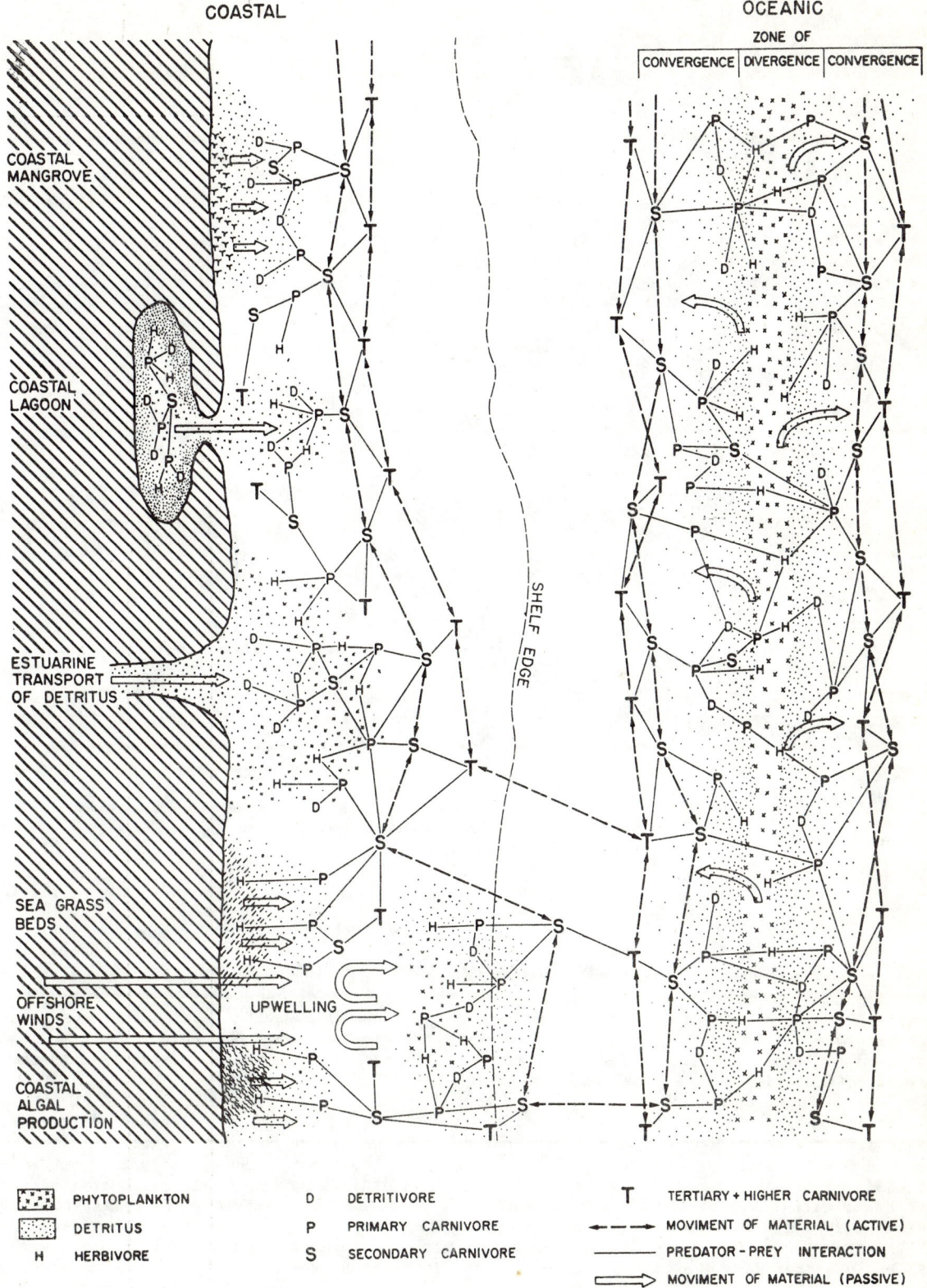

Figure 3 Illustrating the localized nature of centres of coastal and oceanic production, and the lateral diffusion and transmutation of original primary production via food web linkages, and through passive dispersal and migration of food web components. These key areas of production are referred to here as "dissipation structures"

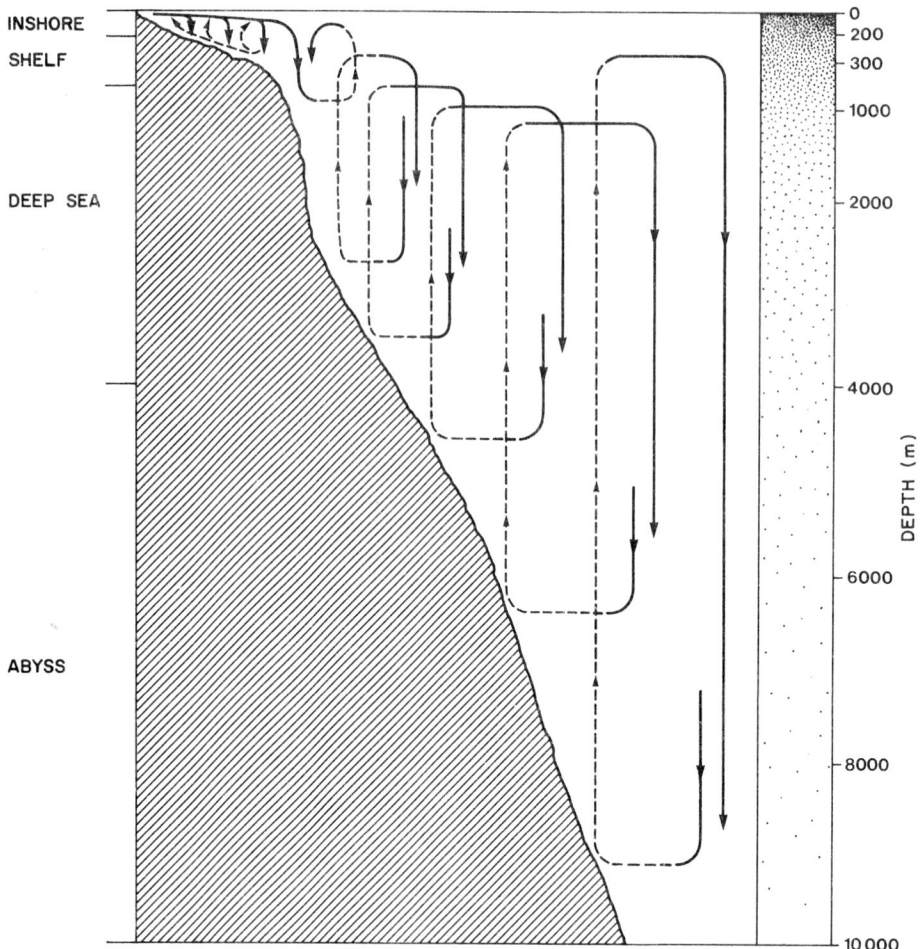

Figure 4   Ladder of vertical and horizontal migration as a means of introducing surface production to the deep sea. Changes in relative concentration of particulate matter with depth in the oceans is shown by stippling in the right column (Modified from Conover, 1978 after Vinogradov, 1962)

In a sense, what is implied here, is what is implicit in the term 'biosphere'; namely that at any point in time, there is only one planetary food web. Derived directly from this abstraction, is the key to working with the food web concept, namely, that those components that regularly or intensively interract, can be extracted from the whole and discussed as semi-autonomous groups or sets. However, it follows that these food webs in the limited sense used here, should properly be considered as 'open': i.e., having some degree of linkage to unspecified components outside those under direct consideration at the time.

In all areas we recognize the distinction between primary use and recycling of waste material; organic matter generated principally by photosynthesis from simple inorganic compounds flows upward (i.e., to 'higher' trophic components) through the food web, but is also cycled back downwards from each food web component; especially in the form of fine organic particles, to re-enter the food web lower down, either via bacterial or fungal action, or directly by detritivores or scavengers. It is prncipally this fact that makes the concept of the trophic level, though still widely used in the literature, very difficult to quantify and verify experimentally. Its use in the present paper is to be interpreted in the qualitative sense, of "distance above the level of primary production".

This dispersal of the original localized production of organic material goes on vertically also (Figure 4), in that the highest production occurs in shallow water and close to shore, and (especially in the tropics), drops off relatively rapidly with depth (Figure 5). Here again, the movement of material into greater depths is facilitated by currents, sedimentation and vertical, diurnal or seasonal migrations, and may pass through several linkage in the food web in the process.

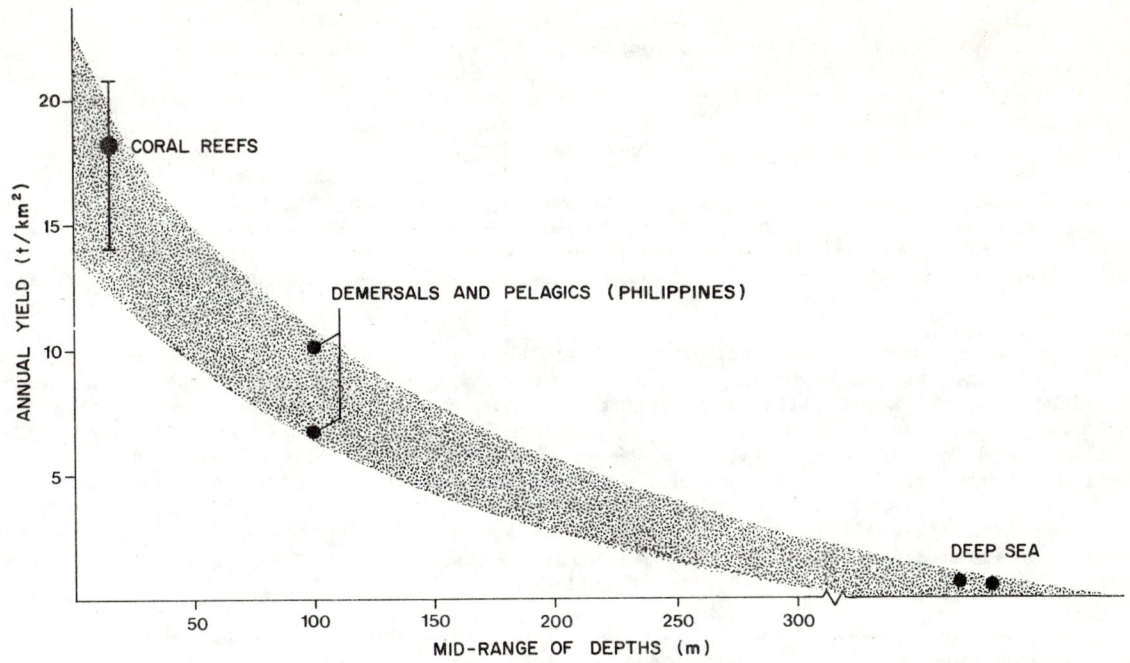

Figure 5 Fish yields per unit area from Philippine waters in relation to depth
(The shaded area is a subjective assessment of possible ranges, not a confidence belt - redrawn from Pauly and Mines, 1982)
(In north temperate waters, the rate of decline in yield with depth is generally much slower)

Those areas of estuaries, lagoons and coastal marshes, between purely freshwater and purely marine environments, represent an important but physiologically difficult environment for most aquatic organisms. Despite the high productivity of most estuarine areas, and their role as nursery grounds, the species diversity of these areas (Figure 6) is generally lower, however, than for either freshwater and fully marine environments, especially in temperate waters. Similar low diversity and high production is also characteristic of (fully marine) centres of upwelling.

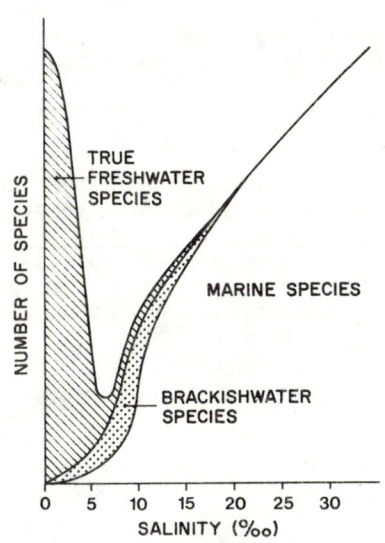

Figure 6 Relationship between salinity and species diversity for largely temperate environments (Redrawn from Pauly and Mines, 1982)

Figure 7 shows the nutrient recycling function occurring in parallel with the primary flow of organic material. This series of feed-back loops for recovery of "waste" organic material must be one of the major stabilizing factors in food webs, since it spreads out short seasonal pulses of production. This is particularly the role of the large number of detrital feeders, especially in the high latitude benthos (where many such species are relatively long-lived, and provide food for many demersal fish on a year-round basis). The time delay and storing of nutrients through time from the short seasonal pulse of planktonic production harvested by the benthos, may be a major factor contributing to the generally greater stability of demersal compared with pelagic fish stocks. The relatively lower abundance of demersals than pelagics in tropical environments, is perhaps in part a consequence of the lower densities of these longer-lived infaunal invertebrates, which in consequence, may play a less important role in buffering production in most tropical systems.

Figure 7 also illustrates that the peaks in production are likely to be smoothed out with distance from the base of the food web, so that the possibility of modelling each component under an equilibrium, or close to equilibrium assumption, is more realistic for the high (apical) components, where fluctuations are less pronounced, than at the base of the food web. In Figure 3 we also see that the dispersal of material from the main centres of production goes on with distance above the base of the food web, and that especially for the larger, more active apical predators (e.g., tunas), the need for active swimming or feeding migrations stems from the greater average distance between the dissipation structures generating an abundance of the larger food particles necessary for their energetic balance. This in turn, increases the metabolic rate of predators, and thus the amount of food they require for metabolism.

Higher-than-average peaks in the primary production in favourable years are passed on to secondary and subsequent trophic levels, smoothed and reduced in amplitude, as well as progressively lagged in time behind the original date of production (Figure 7). To some extent, this is a consequence of the multiple trophic levels occupied by many higher predators. The fact that these may occupy lower trophic levels as larvae or juveniles, means that their recruitment is subject to the same fluctuations as lower components, with which they share the same level in the food chain as larvae. This observation goes a long way to explaining the wide fluctuations in abundance of many marine fish species, even when they are high in the food web as adults. This aspect, together with other socio-economic uncertainties (Figure 20) is one of the inevitable aspects of "boom and bust" that the fishing industry is subject to (e.g., Caddy, 1983; Sharp, Csirke and Garcia, 1983).

Of course, the large seasonal fluctuations illustrated in Figure 7 will be less pronounced outside the arcto-boreal and upwelling regions, but the same general principles probably apply in the tropics although the rate of turnover will be higher, and production and recruitment peaks are likely to be less pronounced.

Another aspect of the transfer of energy is indicated in Figure 7: either material is passed up in the food web by metamorphosis (in the case of species where a transition from low to high trophic levels occurs with individual growth, or ontogeny); or by predation. In the latter case, a considerable proportion of material is lost in the exchange [see PART II (14) for some estimates of this], and in both cases a significant amount is lost in the form of heat (respiration) while performing the work necessary to life; and as excreta. For tunas, and (to a lesser extent, for) other active species (e.g., krill and squid), the energy component used to fuel muscular contractions (especially swimming) uses up a great fraction of the chemical energy in food consumed, leaving less available for growth and reproduction. A significant fraction of energy reserves of marine organisms also enter the food web as that high proportion of generally small particles (eggs, sperm and larvae) that do not survive to maturity.

One aspect of fisheries systems that has become evident to biological oceanographers in recent years, is that the characteristics of biological systems that at first sight would seem to be of primary interest to man as an exploiter, are the biomasses of useable organisms present. In practice, however, the two features that most determine the ability of a system to sustain production for human use, are the fluxes (or flow rates) of materials through the ecosystem and to man; and for (economic) human harvesting, the availability of organisms to harvesting. This latter in turn, is usually dependent on the degree of concentration or aggregation of species.

4. CHANGES IN TROPHIC LEVEL IN THE LIFE HISTORY

Morphologically distinct larval stages with their own environment (niche) and food requirements are well developed in most marine organisms, and in a certain sense, the larval stages and individual development of many species seem to echo at each stage in the life history, the typical trophic interactions of adult organisms of the same size occupying the same environment. Thus, the larval stages of herbivores as well as higher predators, usually start at similar positions in the food web, and apical predators may thus be regarded as moving upwards in the food web in ontogeny

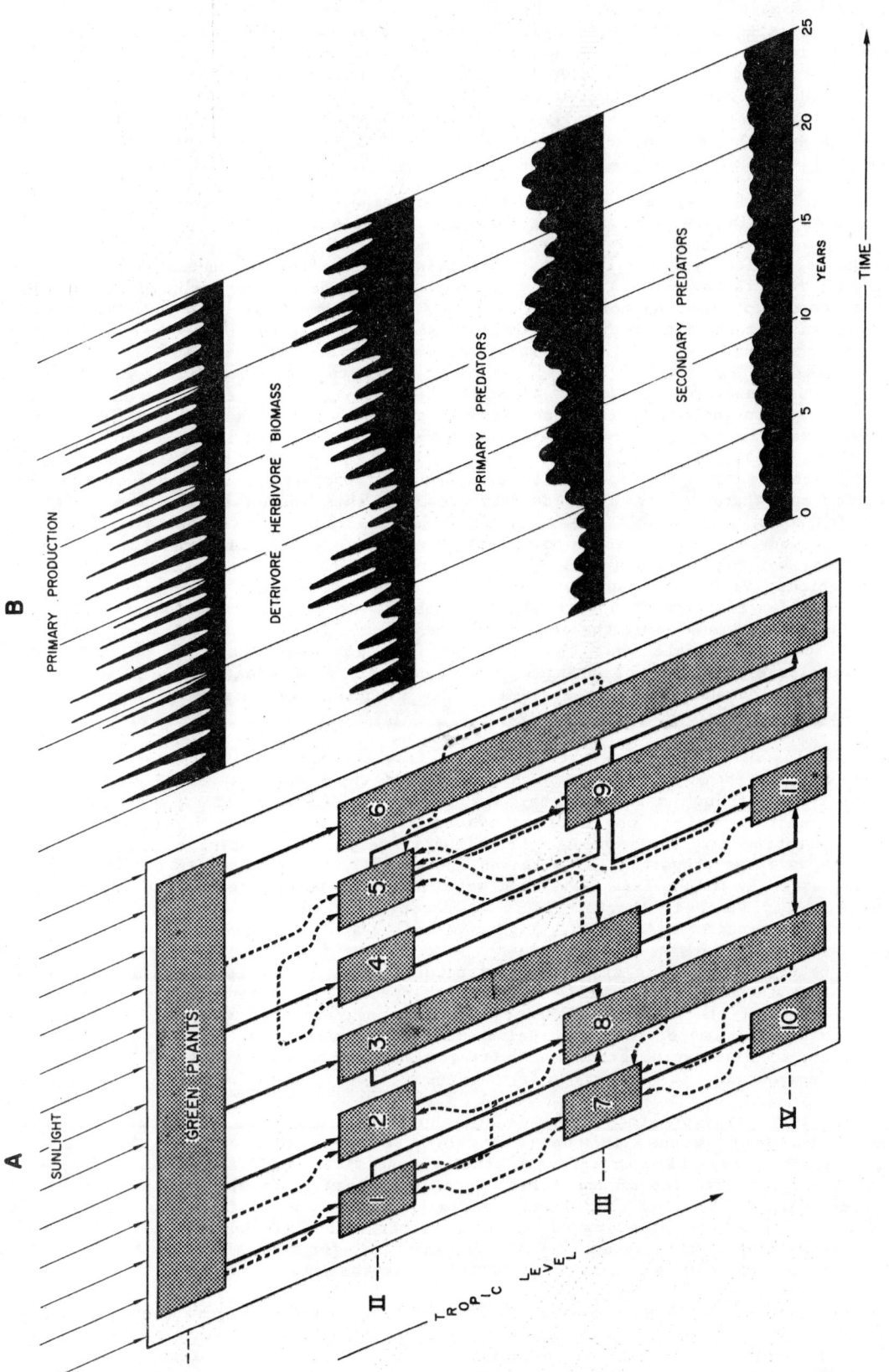

Figure 7 Diagrammatic representation of food webs in time, showing (A) a simple food web with linkages between successive "trophic levels" (straight solid lines), (B) the multiple trophic levels shown by any macroscopic marine organisms, as well as (C) the often "reverse" flow of biomass through detritivore loops (dotted curved lines), (D) the temporal succession of peaks in production which occur on a seasonal and annual basis at different levels in a food web. Particularly for higher latitude systems, these peaks are smoothed out and delayed higher in the food web

(development). This is illustrated in a general way by the elongated "parallelograms" in Figure 7, which show a species spanning several trophic levels, and in Figure 10A which shows that such 'discontinuities' in diet with size are recognizable using statistical criteria. One feature that follows directly from Figure 7 is that since larval stages are usually at a lower trophic level than adults of the same species, they participate in the greater uncertainty and higher temporal variation characteristic of these lower levels: this is probably one of the reasons why recruitment variations are so marked in the sea, even for high level predators. As noted, it makes more sense in this situation, as proposed by Elton (and later by Platt, Mann and Ulanowicz, 1981), to categorize feeding levels by size of predator and preferred size of prey, than in terms of some difficult-to-quantify concept based on knowing the whole feeding history of an organism (Cousins, 1985).

These higher level predators, in common with the other components below them in the food web, will of course remain within the web on dying (through predation, disease, senescence, etc.), whereupon scavengers and bacteria recycle their organic and mineral contents and incorporate them once again into the production cycle. Liquid and particulate waste from living and dead organisms is thus recycled back into the system. This provides another secondary layer of complexity to the food web, and also tends to make the concept of a single species-specific trophic level largely meaningless, since a micro-herbivore or primary predator such as a copepod, may in part be feeding on primary production, and on fine debris resulting from, for example, the predation on a tuna or a whale by pelagic sharks at high trophic levels, and hence share the same trophic level, at least temporarily. The all-pervasive role of marine bacteria which are either in suspension attached to fine particles, or in bottom sediments, in metabolising organics, provides much of the food energy on those particles of material consumed by filter feeders, and must be taken into account here. If we must talk in terms of trophic level, and this is only one of several possible paradigms as discussed later, it seems intuitively correct to assign marine bacteria a low level in the food web, as for plants, and therefore to place detritivores for this reason at the same level as herbivores when calculating mean trophic level (if this exercise is worth attempting). This is a convention that for convenience considers that whatever the source of organics they are metabolising, the food web begins again once the individual identity of an organism is lost on consumption or excretion, and was the approach followed in the International Biological Programme (IBP). From 1964 to 1974 this attempted with mixed success to determine the productivity of all terrestrial and aquatic ecosystems using the concept of the trophic level as a basic tool (Cousins, 1985). Perhaps it was inevitable that this program fell somewhat short of its ambitious objectives; this was largely because of changes in our conception of ecosystem processes. The large amounts of information and insights obtained in the process have, however, played a major role in the way ecology has developed subsequently.

The basic principles of energy conservation and its flow between organisms in food webs are however of more importance than a philosophical discussion of the trophic level of detritivores. It is quite clear that although marine bacteria and organic detritus allow more efficient reutilization of organic material, there is no system capable of circumventing the laws of thermodynamics and inventing perpetual motion by continually recycling the same biomass. At each step in the food web, a large proportion of received energy is converted back to carbon dioxide, inorganic salts and water by metabolism. The food web thus constitutes a so-called "dissipation structure" (Johnson, 1981) which is characteristic of how primary energy, whatever its form, is dissipated from centres of concentration, e.g., Figures 3 and 4, and later chapters.

A food web can be viewed then as a dissipation structure, where energy is passed along trophic linkages from the original point of synthesis in time and space; a considerable proportion being metabolized at each stage of the original solar or chemical energy input, before the remainder is passed on to the next stage. One of the principal themes of this paper will be to explore how the linkages in these marine "dissipation structures" (that also incorporate fisheries), can best be conceptualized and represented.

The above conceptual framework then explains how short- and long-term perturbations to the system will be 'damped out' higher in the food web. It was suggested that "equilibrium conditions", or a close approximation to the same, that can permit "steady state" fisheries, will be more likely to occur near the top of the trophic chain, so long as levels of exploitation are reasonable and controlled. The view that seems to emerge is then of a fisheries system as a 'damped oscillator'; receiving variable inputs in time, rather than a closed system or chemostat, which has been the general view until recently. At the same time, for many fisheries the idea of a "steady state" maintained over the long term is probably not appropriate - see Part II(1).

As a value judgement that will not be universally accepted, we would suggest that a knowledge of the main features of the food web in its simpler, non-quantitative form, is informative and useful; illustrating the possible impacts of exploitation of one or more of its components on others. This is true even if a quantitative evaluation of the impact of man-induced changes, and of the level at which they begin to show up, cannot be arrived at without a far greater amount of information. In some circumstances, it will also be useful to consider subsections of the food web (for example the fish only; or even the adult fish trophic interactions) since as we have already shown, any food web short of the total biosphere, will have undefined linkages. However, reliability and utility of food web applications increase dramatically as quantitative information accrues.

As a final observation on this point, most marine fish species have a much greater diversity of preferred food organisms than is the case for most terrestrial carnivores, except possibly during the critical larval stages of fish, but preferences, for size and often type of food organisms, are very specific (e.g., Figure 10A). One tentative conclusion that arises from this observation is that variation in the spatial distribution and availability of food of a suitable type and size range can surely be demonstrated to be a key cause of population fluctuation in marine ecosystems, but such variations are likely to have their most serious impact on early larval stages. Conversely, a temporary increase in one or more food items is likely to attract or result in population aggregations or increases of mobile or opportunistic predators that will utilize and modify this temporary oversupply (Murdoch, 1969). This may be regarded as one reason why exploitation by man can occur without drastic changes in population structure in some marine ecosystems (see Ursin, 1982).

5. "BOTTOM UP" OR "TOP DOWN" APPROACHES TO ECOSYSTEM ANALYSIS?

One of the arguments that has been a source of continuing controversy for marine ecologists is that between "bottom up" and "top down" approaches in understanding and working with ecosystems. In simplest terms, the "bottom up" strategy argues that understanding, and more importantly, applying information on ecology requires that we first define the physical environment, then the primary production, the production of herbivores, primary to tertiary carnivores etc., in sequence, from the "bottom" of the food web up to its "peak", and the apical predator of prime importance (which now usually includes man). Not surprisingly, some of the main proponents of this school have been biological oceanographers. Implied also in this approach is the concept that food limitation is a controlling factor in biological production. To a certain extent it is possible to make rough predictions of fishery yield from primary production data (e.g., Figure 9). These predictions are rarely accurate enough to have practical significance for fisheries management (although new developments in remote sensing for oceanology are changing this perspective).

The "top down" school has taken the view that since we are more likely to know (and usually need to know) more precisely the biomass of the apical predators in a fishery system, (as well as the predation pressure caused by one of them, man), it makes sense to measure the biomasses and linkages at the upper levels of the food webs more accurately, since these are usually of most immediate value to the fishing industry. Implied also in this approach is the concept that predators "control" the biomass of food webs components below them: (see Figure 1).

Of course any sensible person is going to take both perspectives into account, but the dichotomy between these two approaches, while hopefully now receding, needs to be borne in mind when reading the historical literature.

6. QUESTIONS OF BIOMASS AND SIZE OF ORGANISM

From Figure 7 we see that the biomass first synthesized by plants, which is the basis for marine food webs, moves "upwards" in the food web from herbivores through the successive trophic levels, either by predation or ontogeny. At the same time, production is dispersed spatially, and its peak progressively delayed in time at higher trophic levels. In general, (and there are many individual exceptions, of which macrophyte browsers, some detritus feeders, baleen whales and parasites include notable examples), body dimensions increase as one moves upwards from one trophic level to the next. Also, at the same time (Figure 11), the concentration of individuals per unit volume of ocean falls progressively with size, as does their rate of population increase and hence their turnover rate. The production per unit weight also drops off. This is in accord with common observations, namely that bigger fish are rarer, and that the vast majority of oceanic biomass (and even more so, production) is concentrated at the microscopic end of the size spectrum, and hence is unharvestable. Of course conventional arguments on the simple basis of size can lead one to very wrong conclusions in some circumstances (see, for example, Figure 10B). The clearest example would be that strict application of the "bigger eats smaller..." tendency would not suggest that the great whales should be dependent upon krill or other primary predators; or that tunas would need and/or employ filter feeding apparatus; or explain the observation that adult herring can eat larval cod (in contrast to adult cod eating juvenile and adult herring). As a qualification to the above, even a good knowledge of the trophic interactions between species is unlikely to fully explain population changes, since environmental fluctuations and intensive fishing pressure are likely to be more critical driving variables than trophic interactions at various stages in resource life histories. (See for example the account by Gulland and Garcia (1984) of changes in species dominance in West African fisheries, and Sharp and Csirke (1983) for other world areas). One needs to be careful of generalizing too widely.

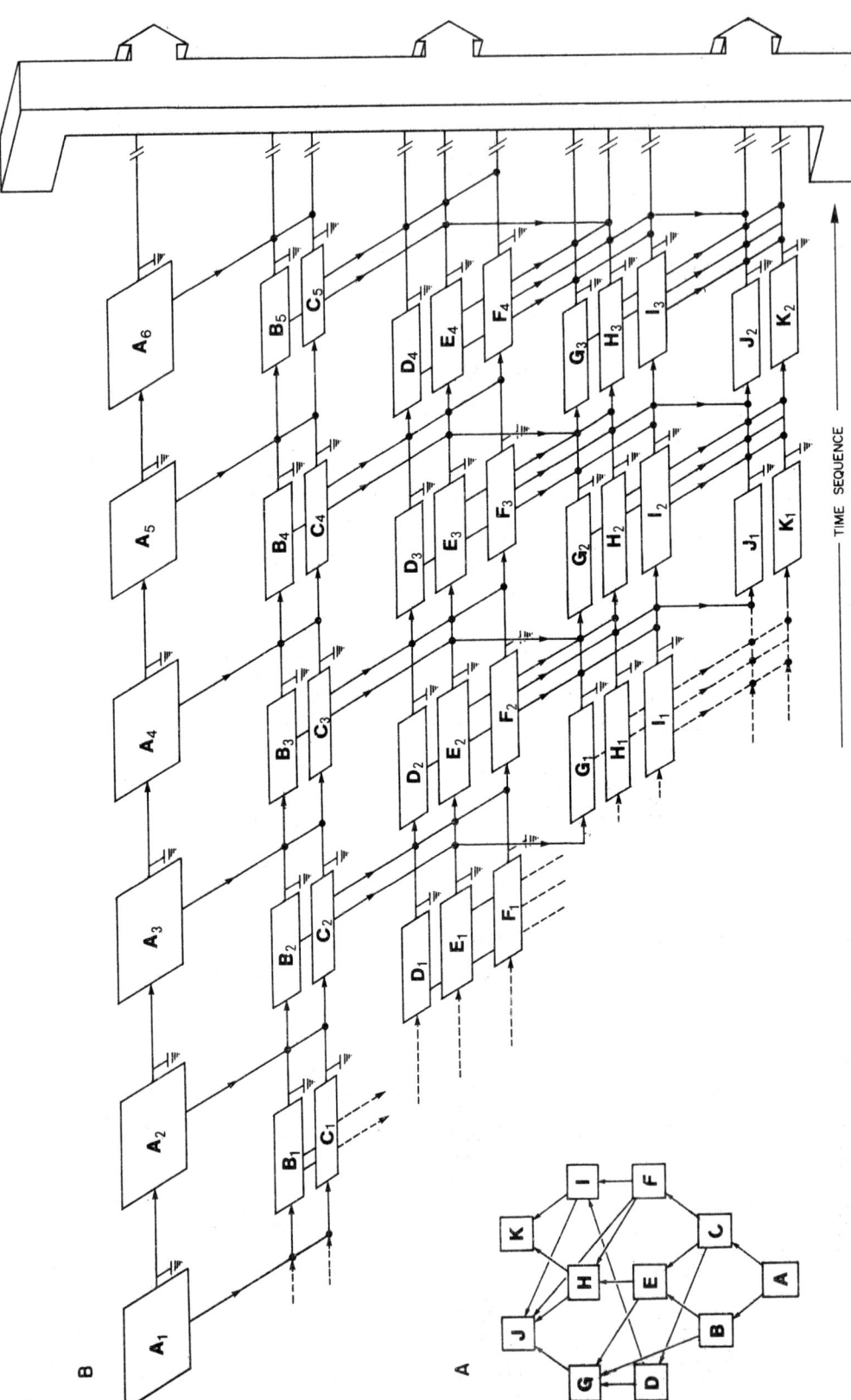

Figure 8   Illustrates how the "static" food web concept (Figure 8A) corresponds in reality to a flow diagram (Figure 8B) in which organic material either preserves its macroscopic identity as a component of the biomass of a given species (by moving to the right in Figure 8B), or is transformed with the inevitable loss of a majority of the accrued production at each on being consumed by a component "higher" in the food web. (The loss in the energy in the form of heat is shown by the vertical arrows "to earth" in Figure 8B). The time sequence of past events has been illustrated by the cursor at the right of the diagram (set in the present), but the time scale of events at all trophic levels is not necessarily constant as shown here. Thus, the subscript numbers $n = 1, 2, 3, \ldots 5$ to the alphabetically labelled food-web components (quadrilaterals or "lozenges") may represent a n'th "pass" of a "package" of material through the food web, i.e., the "mean age" of organic material in days since its original synthesis is greater for higher food-web components than lower levels and the "turnover rate" is slower. Thus, if the extension of the individual "lozenges" is regarded as generation time, these would usually be of longer duration for apical predators (e.g., $J + K$) than for basal components (A). This is not well illustrated in Figure 8B

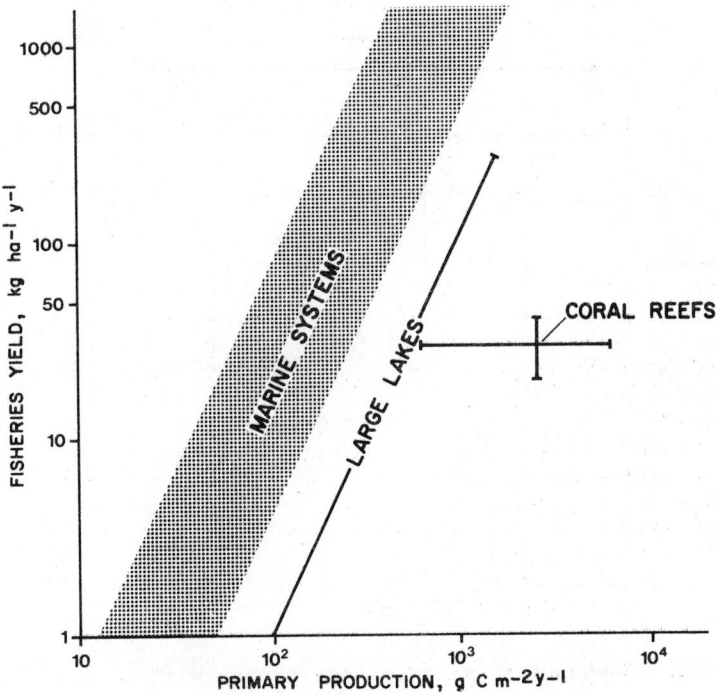

Figure 9 The relationship between fisheries yield and primary production for three main aquatic ecosystems; marine (shaded area), large freshwater lakes (regression line) and coral reefs; (Modified from Nixon, 1982)

7. SOME PROPERTIES OF FOOD WEBS

From modelling studies on food webs carried out to date (e.g., Pimm, 1980; MacDonald, 1983), a number of generalizations seem justified, some of which may have some broad applications to marine ecosystems:

1. Systems with the greatest diversity are likely to be more stable than those with fewer linkages. On the other hand, more diverse systems generally occur in stable environments where they have had a longer evolutionary time to develop, and hence may be less resistant to stresses which exceed in intensity and type those to which they are adapted. Once perturbed, systems with many linkages are also likely to take longer to return to stability.

2. Feeding by most organisms is size specific, and is preferentially directed at the larger food particles available to the organism, and not predominantly to items which are usually separated by more than one step in the food web. However, we should not forget that important exceptions occur: thus, tunas have well-developed gill rakers, and there is a direct correlation of minimum sizes of stomach contents with gill raker gap (Magnuson and Heitz, 1971). (Note also that cases where this generalization is inverted are common, and are usually referred to as parasitism.)

3. Many characteristics of marine systems are broadly speaking, functions of size. Thus r, the intrinsic rate of natural increase (Figures 11A and 12B), and M, the natural mortality rate (Figure 12A), as well as the generation time (Figure 12B), are functions of size; as is the biomass present per unit volume of particles in sea water (Figure 11B), and the organisms' feeding rate (see Figure 67 later on).

4. The fish biomass of temperate regions is made up of relatively few species, and in general the biomass increases, and the production/ biomass ratio and the species diversity decrease, at higher latitudes. One reason suggested for this is that except for monsoon and upwelling areas, production is usually seasonally more uniform within tropical waters, unlike the temperate regions where sharply contrasting seasonal production cycles are the rule. Those

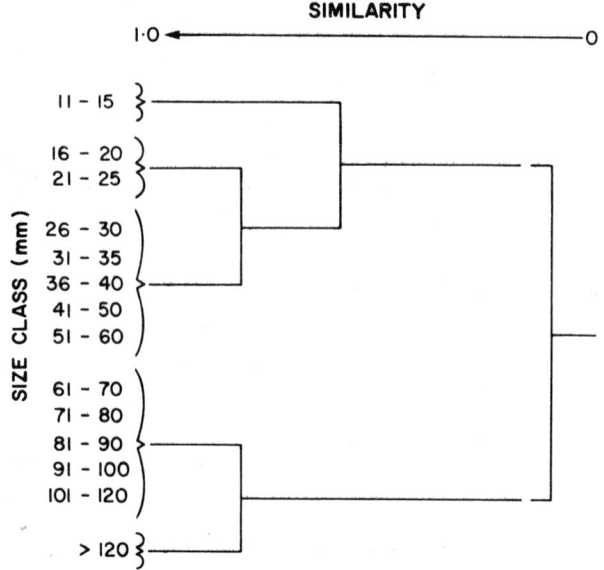

Figure 10A Changes of predator preference with size for species, and variations in predator preference at a given size. Cluster analysis of prey similarity among size classes of pinfish (Lagodon rhomboides) taken in Apalachee Bay from 1971 through 1977 showing that main discontinuities of diet can occur (in this case around 60 mm in size) during ontogeny. (Redrawn from Livingstone, 1982)

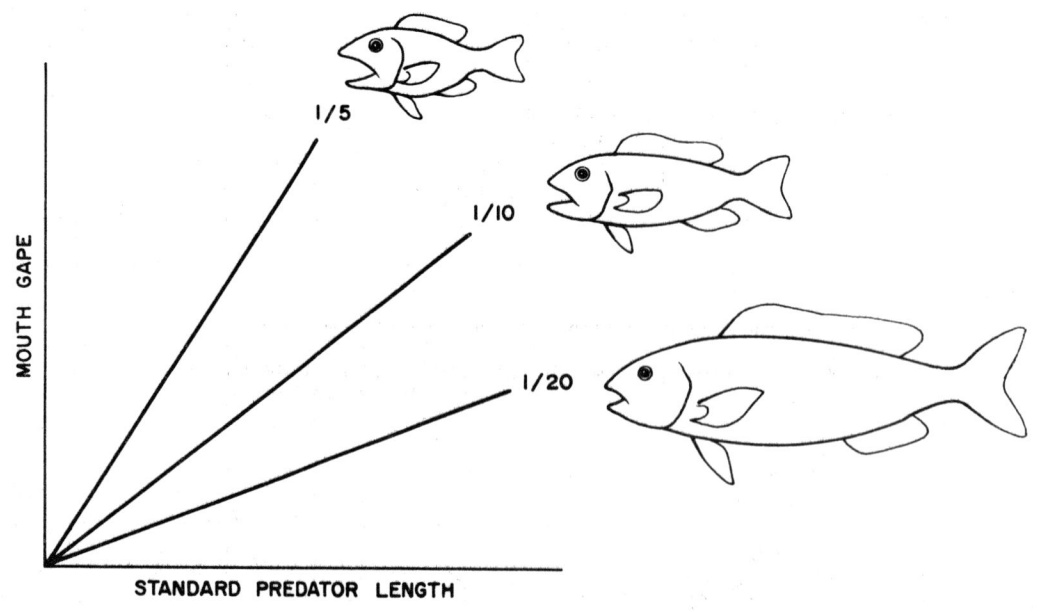

Figure 10B The relationship between the standard length and the width of mouth gape for three different predators with different feeding strategies. (Redrawn from Hoar, Randall and Brett, 1979)

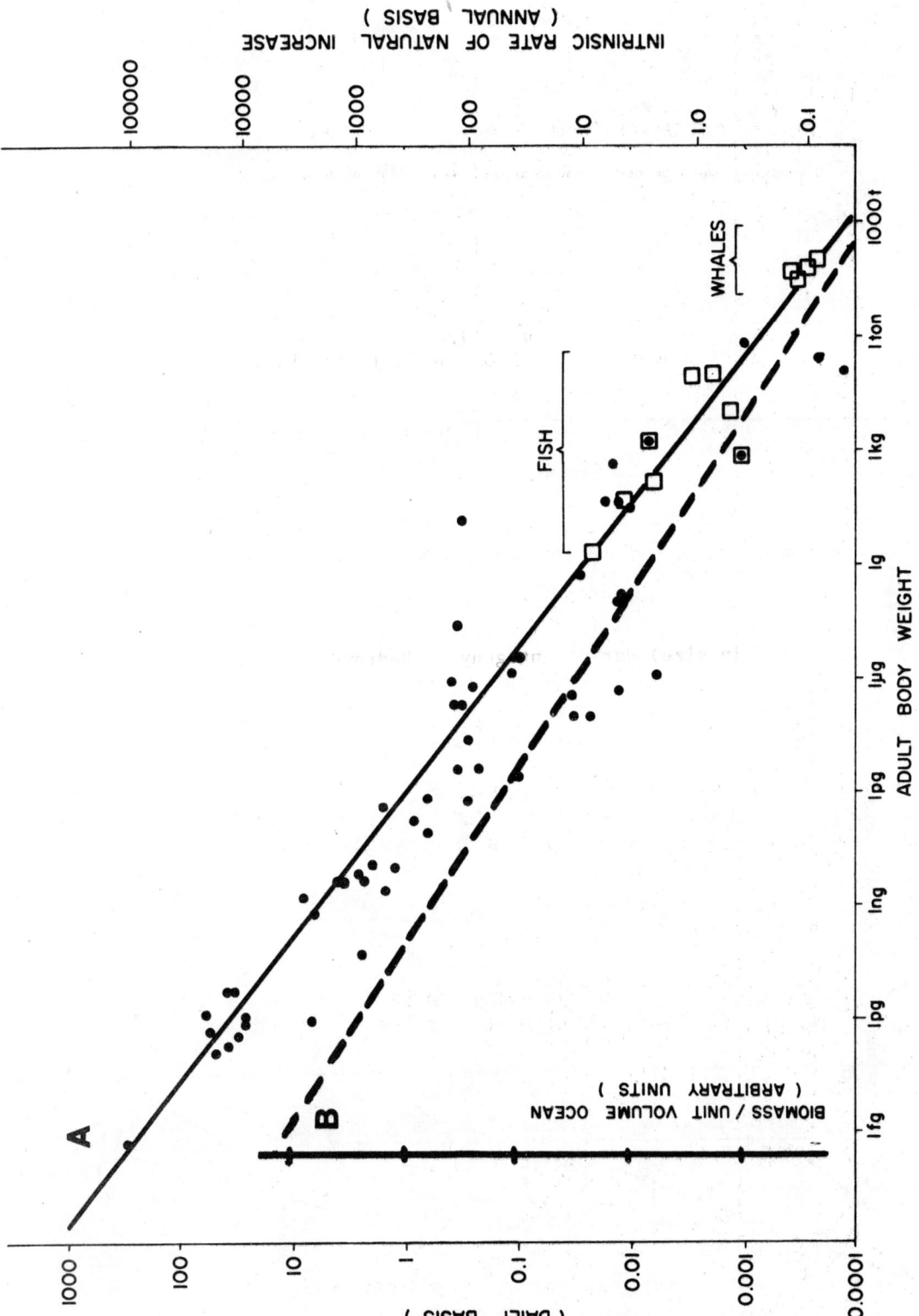

Figure 11 (A) Relationship between the intrinsic rate of natural increase of populations (mainly animals) and their adult body weight (From Pauly, 1984). The regression lines and most data points (black dots) are from Blauenweiss et al., 1978 who also provided the regression equation (see Pauly, 1980)

(B) Relationship between unit weight and concentration of "particles" (including living organisms) in sea water (After Platt and Denman, 1978)

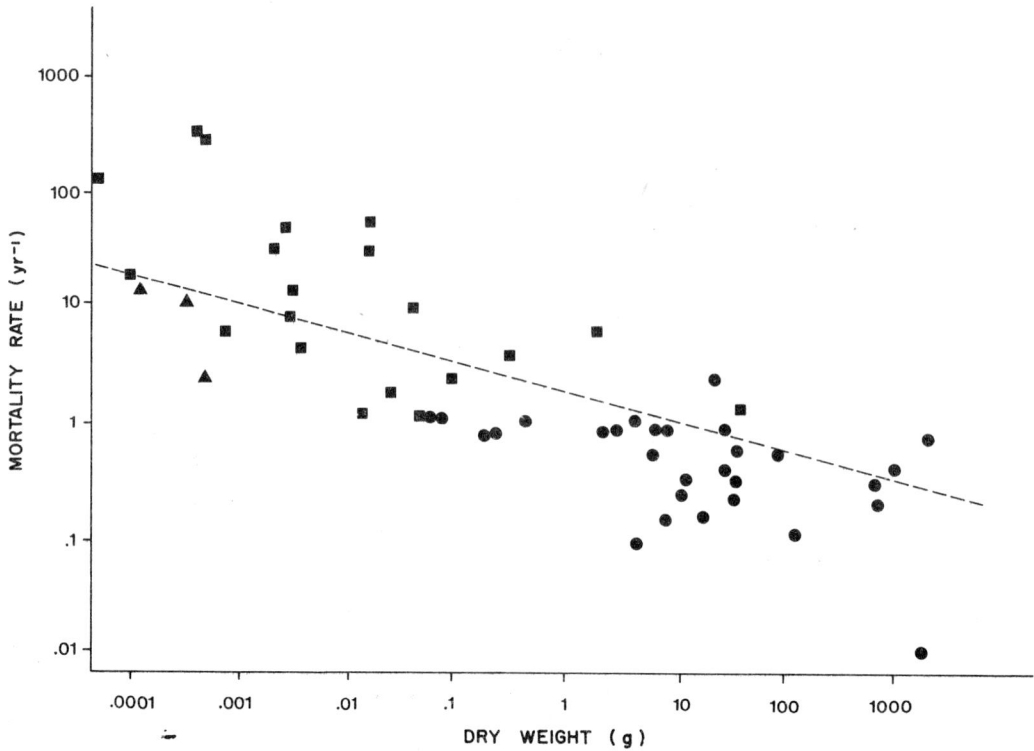

Figure 12A   The natural mortality rate as a function of body weight in marine fish.
(Redrawn from Peterson and Wroblewski, 1984)

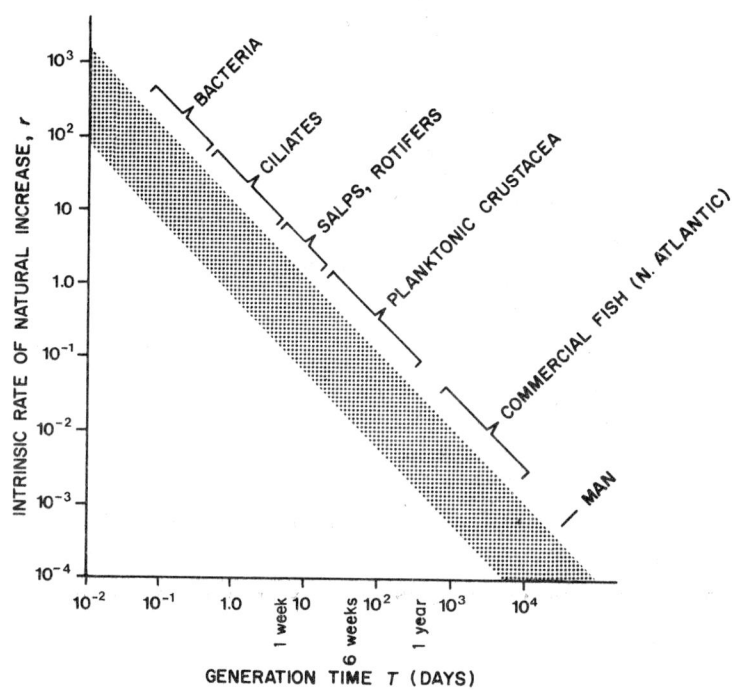

Figure 12B   The intrinsic rate of population increase r,
per day as a function of generation time.
(Redrawn after Heron, 1972; in Anderson, 1981)

species adapted to the "boom or bust" production situation in temperate areas appear to be generalists in their feeding habits, have specific periods in their life history when they feed on a particular prey, are migratory, or use a combination of all three strategies to overcome within-year variations in food availability.

5. A number of key changes in biomass distribution occur with latitude: thus, benthic biomass decreases with depth, and increases at higher latitudes, and the percentage of demersal fish in commercial landings is greater at high latitudes and low temperatures. Jones (1982) derived an empirical expression for this latter relationship from a limited set of data; namely:

$$\text{Percent Demersal} = 100 \exp. - (0.12\ T)$$ where T is the (mean) bottom temperature.

6. Body size at any given trophic level also seems to increase with latitude. Mouth gape/body length ratio similarly tends to increase, and also with depth where species are living under conditions of limited food supply.

As in all natural systems, many exception to such 'rules' apply; and ecosystems and their components require individual consideration, before generalizing too widely.

8. ZOOGEOGRAPHY AND LIFE HISTORY STRATEGIES

Apparently the higher latitudes are difficult habitats to colonize, except for very specifically adapted, arcto-boreal or polar species, implying that these species are preadapted to longer-term seasonal deprivations and longer generation times than most tropical forms. The large numbers of species in tropical systems are associated with a patchy distribution of localized productive areas, and numerous micro-habitats in time and space. The survival "windows" in the early life history (Bakun et al., 1980), may be more nearly randomly distributed in time and space with respect to the range of reproductive adults in these populations, unlike the situation that leads to the synchronized "blooms" and "recessions" observed for temperate species. In general, the occurrence of multiple age groups and/or multiple spawnings are safety measures to ensure continuity of the species in face of quite frequent unfavourable years. This "safety margin" is taken advantage of (but should not be abused) by commercial fishery systems, that in effect use the limited 'elasticity' of natural systems, and harvest it as so-called "surplus production".

The simplest systems, say for example Arctic lakes (Johnson, 1981), are highly seasonal, relatively efficient, and very stable over the long term, and the perturbations here are induced predominantly by energy input (i.e., insolation, wind) on a strongly seasonal cycle. Here, food webs consist of few species. In the sub-arctic ocean there are more species, but still far fewer than in the temperate and in particular the highly complex tropical oceans. As we move equatorward, the influences of local geography, seasonal and climatically induced energy inputs, and the numbers of nomadic apex predators, increase. This implies that closer to the equator, the systems and overlapped habitats become more complex, and less predictable in local species composition and local abundances, and therefore less stable on local geographic and short-term bases. A good review of the role of hydrographic and spatial factors in speciation of marine organisms is given in Sinclair (in press). However, the wider-ranging the species (i.e., broader the habitat), the less likely the species as a whole is to be subject to great fluctuations (see Sharp, 1980b for further discussion of fluctuations on a large versus local scale).

9. LIMITATIONS ON FOOD WEBS IMPOSED BY AVAILABLE SURFACES

Whereas a tropical reef community consists of numerous species, all with rather specific physical and trophic niches (e.g., Randall, 1967), these are usually relatively volatile in regard to both local distribution and recruitment, although it is through continuous small cataclysmic removals and simultaneous recolonization and re-establishment over a mosaic of small potential habitats, that they persist at all. So much of the complexity of local reef ecosystems tends to be damped out as one moves up-scale geographically to focus on the whole zoogeographical area (species range) occupied by individual species.

We have already briefly discussed the importance of the differing physical scales of water movement for pelagic food chains, from small-scale turbulence and their impact on larval life histories, to local areas of larval retention and their relevance in determining potential stock size in pelagic fish, as well as the importance of large-scale water movements and boundaries for transport and production processes (e.g., upwellings). A somewhat analogous approach to describing the physical substrate, and hence its capacity for producing or supporting useful production in demersal fish and invertebrate food webs, is becoming of practical interest. It is clear that although the level of primary production is important, the nature and physical dimensions of irregularities in bottom substrates is of importance for supporting important concentrations of commercial-sized species. Two examples illustrate this theme: (a) coastal fishermen in many areas of the world recognize that areas of "live bottom" (Powles and Barans, 1980), that is, rock outcrops and coral reefs (not to mention wrecks, artificial reefs and offshore oil rigs), attract a higher density of commercial-sized species than surrounding flat bottom areas, although the area of foraging of these fish will often be in other adjacent areas; (b) in the contrary sense, some areas of admittedly high production, such as turtle grass beds on flat sand bottom in the tropics are of great local importance for juvenile fish and shellfish resources, but much of the production here flows into food chains which do not seem to directly support high resident densities of large demersal fish of a size attracting commercial fisheries, even though a variety of commercial species migrate or forage through these areas. Apparently such areas, though productive, do not provide adequate 'cover' for most larger species.

These two examples, although qualitative, prompt the observation that survival of demersal species, especially in the absence of "cover", often depends on specific behavioural adaptations (e.g., burrowing in flatfish, or massive shells in gastropods such as Strombus that inhabit turtle-grass beds). The diversity of fish species and their size ranges which are associated with coral reefs, although related to the productivity of the system, is presumably also dependant on the degree of physical "dissection" of the hard substrate, or the presence of physical niches of appropriate sizes for shelter from predators, as well as providing substrates for the production of fauna and flora. Both xoanthellae (algal cells in the tissues of many corals), but more importantly, fine "algal turf" growing on the coralline outcrop (Johannes, 1972), form much of the basis for reef production in generally unproductive areas, and require hard substrates for their development.

The often-heard criticism of artificial reefs, that they increase vulnerability by increasing aggregation, but do not increase overall production, again seems to be a function of scale: the use of "brush parks" in lakes and coastal lagoons in various parts of the world, seem to show unambiguously that commercial yield can be considerably increased in this fashion (see Kapetsky, 1981), though this effect seems to depend on large-scale manipulations of the habitat that are difficult and costly in the more extensive marine environment.

A concept that appears to be emerging with potential for better understanding of this phenomenon, is that of the fractal nature of substrates or coastlines (e.g., Mandelbrot, 1982). A fractal surface can be roughly described as one whose surface area increases in size inversely with the unit of measurement (Figure 13). Most rough or dissected surfaces and contours (e.g., coastlines, Mandelbrot, 1982) show this feature, which in effect means that there is a greater surface area (and number of "niches" in the more restricted physical sense), available for smaller organisms than larger ones (Morse et al., 1985); a view that seems in accord with the "pyramid of numbers" concept of Elton, described earlier. Figure 13 illustrates one of the corollary concepts, namely that for a fractal surface, the number of crevices (and hence the number of organisms) decreases in a logarithmic fashion with size of measurement unit: a phenomenon that seems to control the size spectrum of organisms (including fish) present on coral reefs. Smith (1975) spells out some of the features in play on a Bahamian coral reef:

"First, large fish cannot fit into small holes and are necessarily restricted to larger shelter sites. Secondly, large fish are not available to predators with small mouths, and finally small fish are limited to small prey. The converse of these statements do not apply, or apply only in varying degrees. Small fish fit in larger holes, but in doing so would be more exposed, and lose the advantage of being small". He further addresses the limitation thus imposed on the food web by the physical environment, including the size spectrum, (Figure 14), and the "absolute individual size limitation" thus imposed - otherwise referred to as the L_{∞} parameter of the von Bertalanffy growth curve: (i.e., the low maximum size reached by many reef fish and crustaceans is a function of crevice size).

Where the implications of the fractal nature of natural substrates have been investigated in more detail, as for terrestrial arthropods (Morse et al., 1985), there is shown a general correspondence to the prediction (Figure 14) of a linear relationship between the logarithms of number of individuals and their body lengths. This has also been predicted by Platt and Denman (1978) from different theoretical considerations for the number and size of "particles" (including living organisms) in sea water (Figure 11B). This "ideal" relationship does not of course apply in all situations, and Bradbury, Reichelt and Green, 1984 recognized that for many natural systems, such as coral reefs, it is precisely the discontinuities in the plot of number of physical niches against size that could impose "bottlenecks" to recruitment processes for crevice dwellers, such as many commercial decapod crustaceans (Caddy, in press). The practical relevance of these concepts to the concepts of habitat enhancement or development seems evident, and there is a growing interest in the application of "artificial reefs" whose effectiveness can be judged by how cheaply they provide niches and growing surfaces of the desired dimensions.

10. FISHERIES ECOLOGY AND ASSESSMENT OF RESOURCES

As noted earlier, the main function of this document is to familiarize the reader with some of the key contemporary themes in the field of fisheries ecology, and to serve as an introduction to the relevant literature. We use "contemporary" in context rather than "modern" or "recent", simply

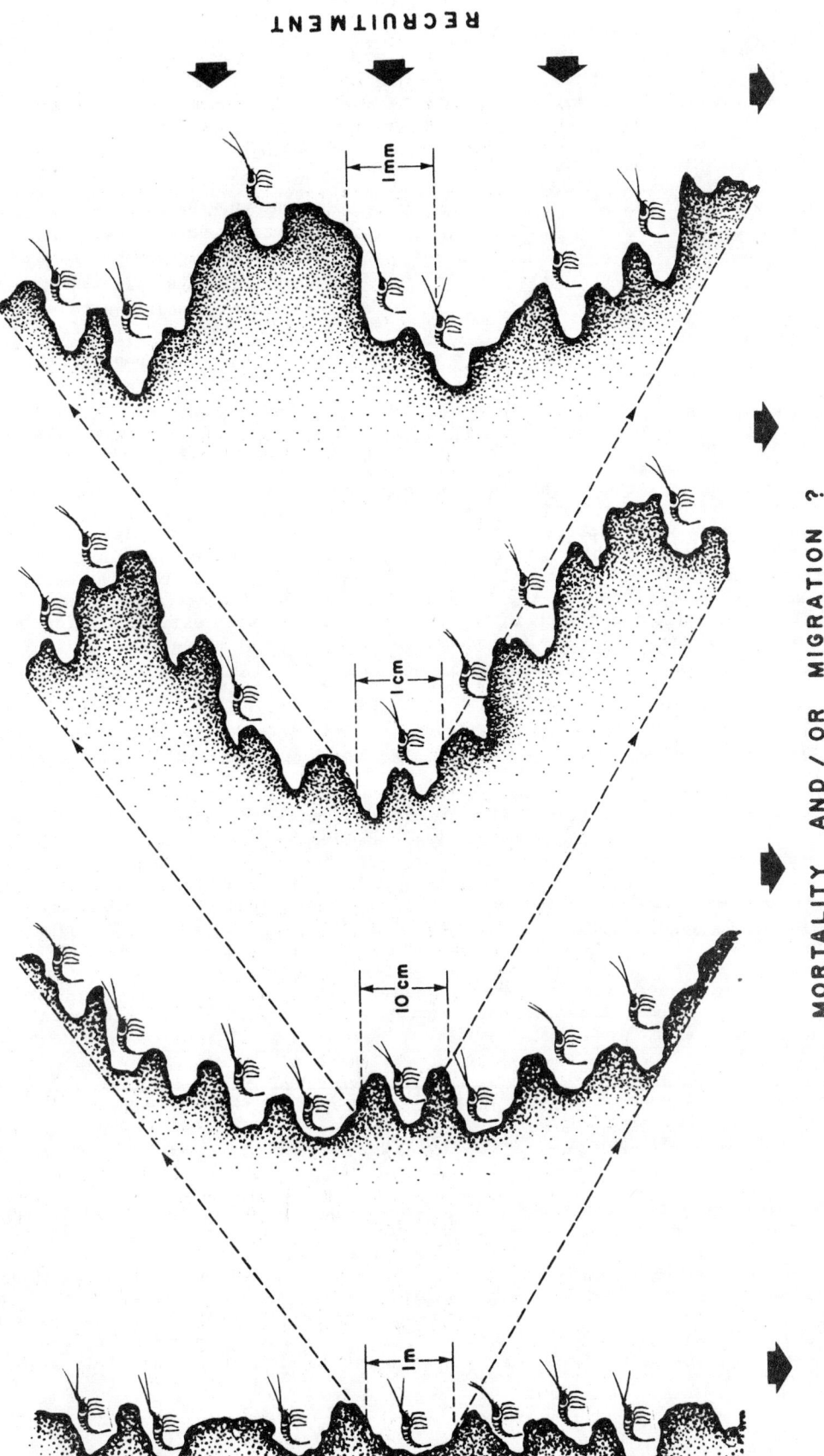

Figure 13 Illustrating four progressive scales of magnification of a fractal surface (right to left), and how crevice-dwelling organisms (e.g., lobsters, most coral reef fish), face progressive limitations in numbers of physical niches available as they increase in size

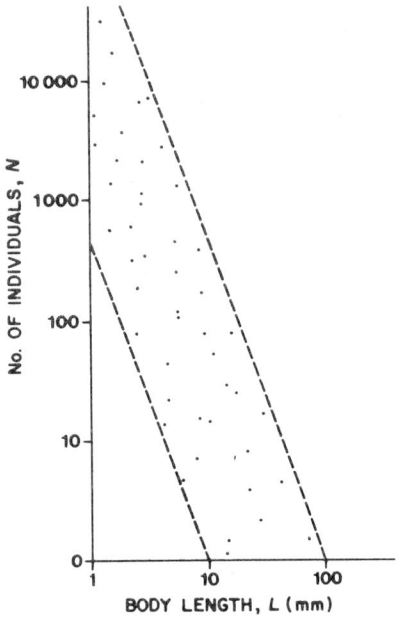

Figure 14 Numbers of arthropods present with size on terrestrial vegetation, showing the fractal nature of vegetation, (i.e., the logarithm of number of physical niches decreases linearly with size). Similar relationships have been postulated for coral reefs and mangrove communities

because the current problems of fisheries biology and their relationship to oceanography and marine ecology, had been preoccupations of scientists from northern latitudes many years before the single species numerical procedures were amplified in the work of Baranov and later Beverton and Holt (1957) to their current position of high profile. These ecological interactions in fisheries are still major preoccupations today. In fact, historically speaking, it may be that the half century or more of ocean science, particularly in the North Atlantic and Pacific, that preceded the first attempts to directly control fishing effort in the 1950's to 1960's, were essential before quantitative assessment methods as applied to sea fisheries, had any chance of success.

Despite this, even in the northern latitudes, the number of consistently well-managed fisheries over the last few decades is generally outweighed by the number where stock collapse, reversible or otherwise, has occurred, not always due to human intervention (see e.g. Caddy and Gulland, 1983). Studies of the ecological interrelationships among commercial species, as well as with their physical environment, are now underway in several areas of the world's oceans, and they are likely to have a significant impact on the way that assessment and management is carried out in the future. Experience with coordinating individual single species assessments, shows that this latter approach is likely to result in the sums of single species MSYs (Maximum Sustainable Yields) theoretically exceeding the whole-systems potential, thereby ensuring over-fishing of these multi-species communities if an attempt is made to fish each species at its individual MSY. It is clear, however, that the development of complex models and computer simulations that do take such species interactions into account, require ever-greater amounts of detailed information and data collection; hence larger staffs of fisheries scientists and administrators, and the costs of such approaches, if taken to extremes, are in danger of exceeding a reasonable fraction of the net income to society from the fishery itself! What is needed are simple, efficient and inexpensive monitoring and management systems, based on the fundamental properties of the biological system being managed, rather than, as to date, just a series of separate approaches to each single stock.

To date despite a growing interest in this theme (e.g., Pauly and Murphy, 1982, May 1984), such a generally applicable theory is not yet available, which means that although species interactions are known to be real, they cannot always be quantified, much less predicted. In the circumstances, a good deal of ecologically-informed common sense is required from fisheries administrators as well as fishery biologists.

Extension of stock monitoring and assessment principles to fisheries of developing countries cannot depend on a similar long historical development of scientific "groundwork" to guide fisheries scientists and managers, since such a background usually does not yet exist, and there are dangers in the wholesale transplantation of inapplicable methodologies from higher latitudes to these less well known tropical stocks (see e.g., Pauly, 1979). At the same time, hasty interpretations from inadequate data will inevitably need to be modified in the light of the systematic, long-term study that will prove to be necessary. It is particularly important therefore that scientists and managers review what is already known about local systems and their properties before descending to specifics in relation to one species or one fishery, and at least encourage some broad-perspective investigation of the local marine systems on which their fisheries depend. In this context, our text does not present a specific set of research strategies. We do suggest, however, that early on in a resource investigation, cataloguing and mapping of resources on an area-season basis, and developing a basic description of the oceanography of the region, as well as of the likely basic fisheries interactions going on there, will be highly desirable bases from which to begin any scientific management of marine resources. Figure 15 presents an example of such an integrated framework for a fisheries investigation that takes most of the types of information needs discussed in this document into account, and could provide the framework to be adapted more specifically to the local situation.

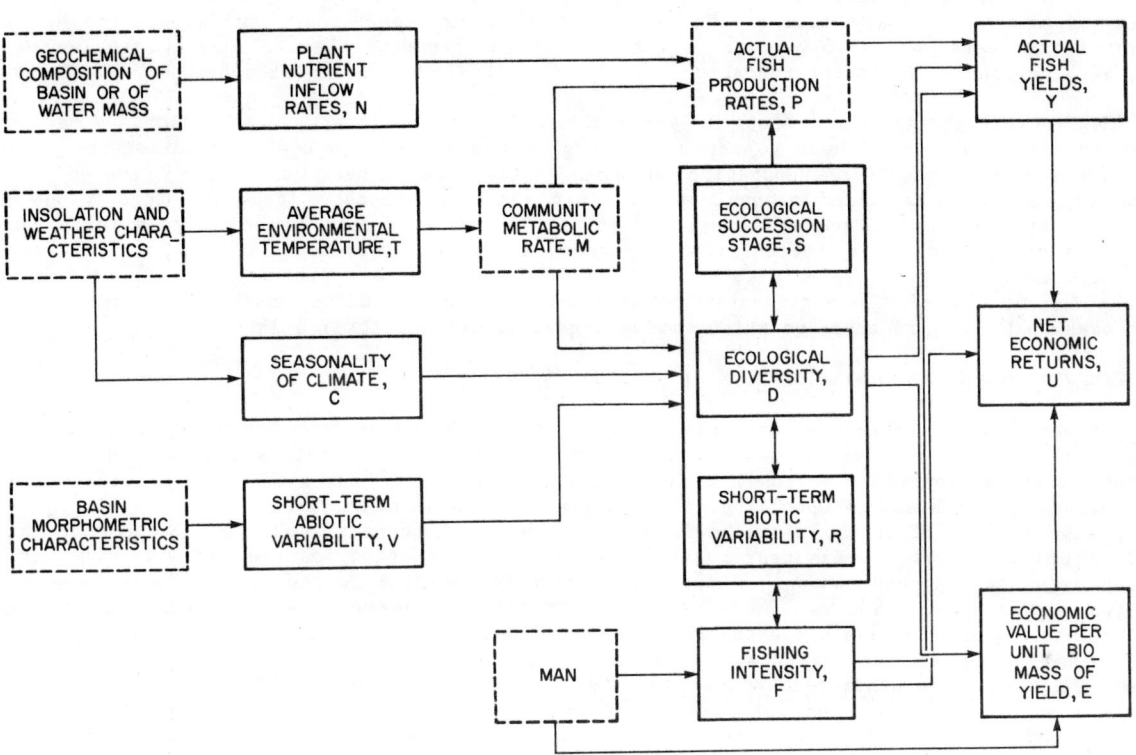

Figure 15 Schematic model for a broad-based ecological and fishery investigation of an exploited natural system   (From Regier and Henderson, 1973)

PART II   ELABORATION AND DEVELOPMENT OF SOME KEY CONCEPTS RELEVANT TO ECOSYSTEM MANAGEMENT

The fact that in harvesting ecosystems we are tapping into a flow of energy as opposed to a static pool is evident, and Golley (1972) notes that:

"The living portion of the planetary ecologicl system requires energy to maintain the thermodynamically unstable condition of life."

and clearly ecosystems thrive where there is a high flux of energy through the environment. Thus Golley notes:

"It is well known that the energy which drives the ecosystems on the plant Earth comes from the sun. At the outer limits of our atmosphere, 1.94 gram calories of solar energy per square centimeter are received per minute. Not all of this energy reaches the surface of the earth; 35 percent is reflected and 17.5 percent is absorbed by the atmosphere and clouds. This leaves only about 47.5 percent of the solar radiation effective at the level of the biosphere.

"How is this energy used? About 30 percent is reflected as long wave radiation, 49 percent is exchanged by conduction of heat to the air through movement of wind over the surface of land and sea. The energy expended in evaporation and condensation and in heat conduction drives the circulation of the atmosphere and oceans, which truly tie the biosphere together into a one-world ecological system."... "Atmospheric circulation acting on the oceans produces the water movements that erode coastlines. Evaporation from the oceans drives the hydrologic cycle, in which water is evaporated from the oceans, precipitated onto the land and returns to the oceans in rivers, effecting the downcutting of the continents in the process. These are the patterns of wind and water which make up the environments in which living organisms act out their existence."

1. TEMPORAL CHANGES AND STABILITY OF FISHERIES SYSTEMS

The major source of energy that drives fisheries systems comes from fixation of carbon dioxide into organic compounds by marine plants, and as we have noted, this process tends to show peak production in rather localized regions, such as (e.g., Figure 3) in areas of high vertical mixing of water masses; i.e., areas of upwelling, in estuaries and over the continental shelves of high latitudes; also further offshore, along horizontal and vertical boundaries between ocean current systems. Highly productive areas nearer inshore in addition to estuaries are lagoons, salt marshes, seaweed beds and mangrove swamps. Dispersal of organic material in the form of organic detritus, plankton, invertebrate biomass, and (actively) fish from these centres of production goes on in space and time from the original date and location of organic synthesis. This occurs both passively by means of eddies and currents, and more actively by means of migration of living organisms: all of the time the synthesized organic material forming the bodies of plants and animals, moves "upwards" in the food web, and at each step towards the size range of interest for commercial exploitation, loses roughly nine-tenths of its biomass and chemical energy in the form of metabolic heat, work, excreta (soluble organics), and detritus. As illustrated earlier, some of the latter two categories are cycled back into the food web by means of marine bacteria and other unicellular organisms, scavengers and detritivores (e.g, filter feeders). Despite what is believed to be the generally low productivity of clear water tropical areas, coral reefs also act as centres for production, concentration and elaboration of organic material, although as for other complex ecosystems, there is a great deal of recycling of the original synthesized material within the system.

The limits to transport processes and in particular, nutrient exchange, in stratified oceanic water between bottom water, rich in nutrient salts, and surface waters where such minerals are severely depleted, must largely account for the generally lower productivity and standing stocks in central oceanic areas, distant from the main centres of production. This is especially true for offshore waters in the tropics, where stable thermoclines may restrict high production to near-shore areas. Peak production elsewhere is usually rather restricted in time, in most regions of upwelling, in north and south temperate phytoplankton blooms, and in seaweed and eel grass beds. This means that the herbivores feeding on plant material, and the primary carnivores feeding on them in turn, tend to be often (but not always) small, short-lived (annual species) with big population blooms and crashes on a seasonal basis. This seasonal peak in production is also true in the tropics and sub-tropics in those areas where seasonal monsoons drive upwelling zones; for example, in the northern Arabian Sea and off Peru, but not so obvious in other tropical areas where a more consistent level of production continues throughout the year.

Generally speaking then, seasonal effects apply throughout the world's oceans, even in clear waters tropical areas and in the arctic, where fluctuations in water temperatures are less marked seasonally. Food web studies (cf. Figure 16), need to take these short-term fluctuations into account. Many of these seasonal changes in diet are associated with transient (short-lived or migratory) species that are frequently voracious feeders, and can significantly affect directions and quantities of energy flow in the ecosystem when present: (e.g., Figure 24). Many fisheries can

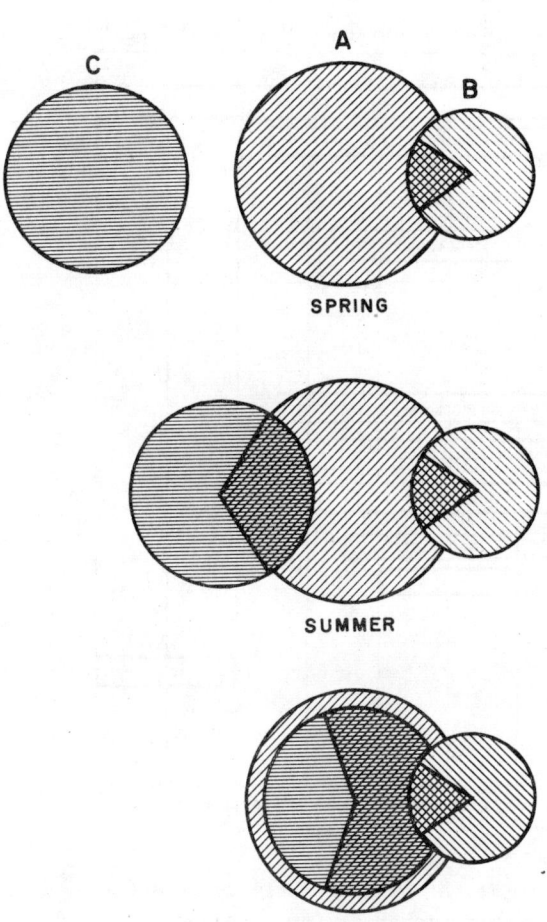

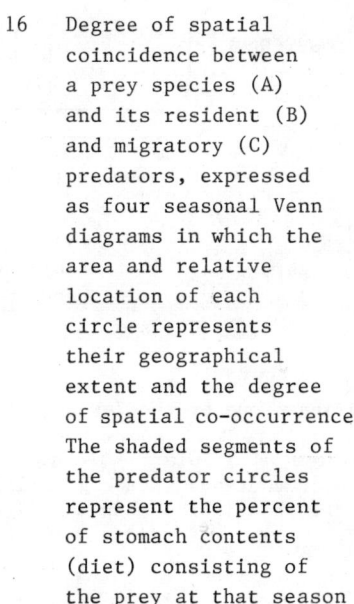

Figure 16 Degree of spatial coincidence between a prey species (A) and its resident (B) and migratory (C) predators, expressed as four seasonal Venn diagrams in which the area and relative location of each circle represents their geographical extent and the degree of spatial co-occurrence. The shaded segments of the predator circles represent the percent of stomach contents (diet) consisting of the prey at that season of the year

also be viewed as seasonal predators and show a strong seasonality of activity, which is generally related to the temporal availability of different species to capture (e.g., Figure 17), as well as to the often associated fluctuations in market price.

As we have noted, one way to visualize the flow of food energy in time and space is to view the food web as a dissipation structure, where input energy from the sun, now in chemical form, moves up through the food web, at each step losing a large fraction as heat in the course of work performed by individual organisms in carrying out the various metabolic processes necessary for the maintenance of life. At the same time (and this is a function of the size scale of human harvesting) the value of the individual energy package generally increases, so that a small fraction of the original (low unit value plant material) becomes often high value fish tissue (e.g., tuna, codfish, etc.), which is energy-rich, harvestable in a cost-effective way, and preferred for human consumption, and usually commands a higher market price.

One of the features of natural systems that is now becoming better understood, is the degree to which fluctuations over the medium to long term in climatic factors are a normal feature of global environments and act as forcing functions in determining annual levels of recruitment, production and annual yield, acting through the level of primary production of the ecosystem.

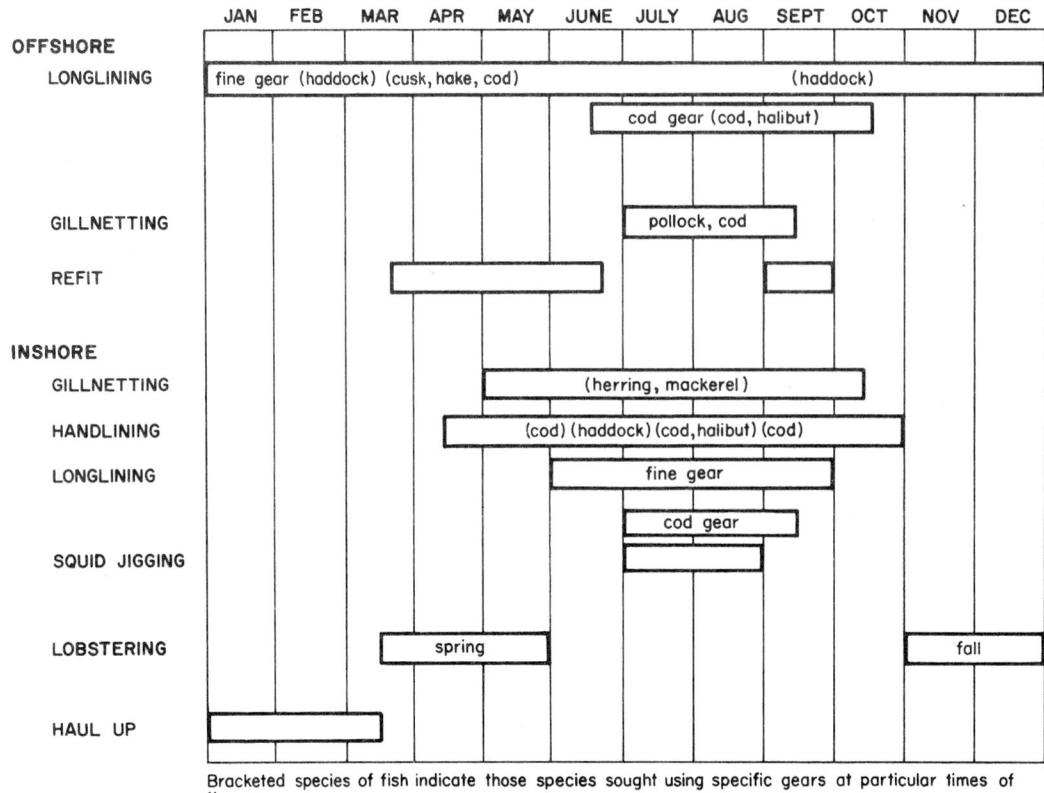

Figure 17 Illustrating the seasonal distribution of fishing activity by species, type of boat and gear, typical of many fisheries (From Lamson and Hanson, 1984)

Figure 19 shows one of the few series of oceanographic data that began to be collected prior to the Second World War, that shows the order of magnitude of year-to-year changes associated with fisheries production. Solar, volcanic, as well as human impacts on the climate are all involved in long-term fluctuations (see e.g., Gilliland 1982). These events may be synchronized over very wide areas (e.g., the El Niño phenomenon, which appears associated to climatic changes on a Pacific-wide basis, and is now referred to as ENSO: the El Niño/Southern Oscillation phenomenon) – (Rasmusson, 1984).

In placing any natural system along the stability scale from close to equilibrium to wild fluctuations or periodic population crashes, the time scale of production is evidently very important: the flow of energy through an ecological system, especially close to the base of the food web is unlikely to closely approximate stability if production is seasonal or even worse, erratic. The chances of managing a system are likely to be improved if there is a relatively continuous stable production, and fortunately even in very seasonal (but regularly fluctuating) systems, higher level organisms have often evolved life history mechanisms to ensure that their production 'evens out' somewhat the seasonal cycle by adjusting their diet at different times of year, or during different life history stages, to seasonal changes in prey abundance, or by migration to locate other areas of peak production.

In fact, a whole new group of life history strategies come into play once evolution of multi-age group forms is widespread, and clearly, new strategies for dealing with the "feast and famine" situation caused by seasonal productions cycles had to be evolved, and as indicated above, these include active swimming, seasonal migration, separate feeding and spawning areas, 'omnivory' (feeding at more than one trophic level), and the whole range of associated morphological and behavioural characteristics that make up contemporary fish species.

The big evolutionary advantage of a longer life span that results when the above problems are solved, is that in evolutionary time, these species have survived instabilities in the environments by adjusting the life span to be roughly equal to the average interval between favourable spawning years, so as to survive a poor spawning year that might otherwise be disastrous for localized

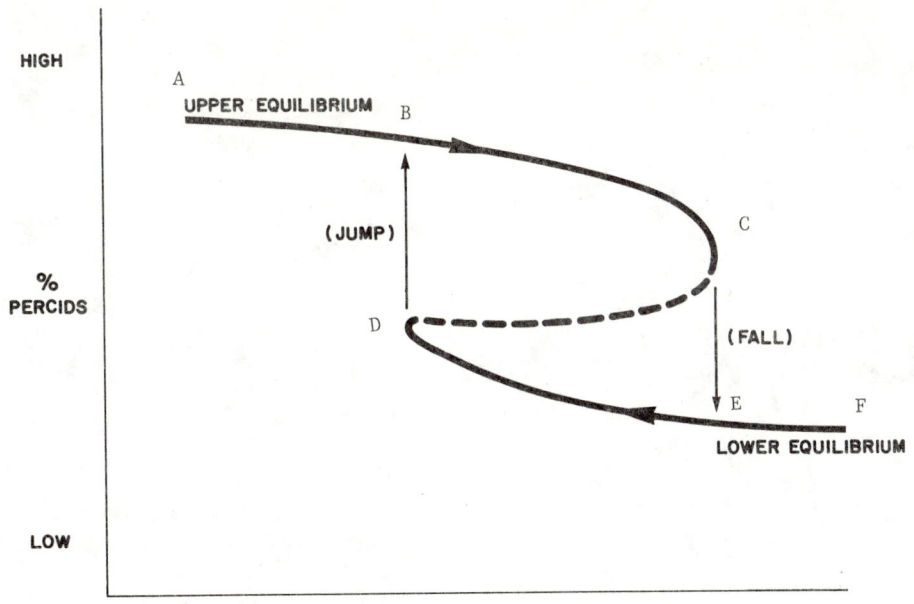

Figure 18   Not all species compositions in a marine community change gradually with fishing (or environmental change), but can on occasions change catastrophically. This diagram, freely adapted from Kerr (1977), shows how the abundance of a species (here a Perciform fish), can change abruptly with various types of environmental stress. Thus, an increase in effort (or a change in some environmental variable) from A to F might lead to species abundance following trajectory A-B-C with an abrupt fall to E-F, but in the opposite direction return via F-E-D-B-A (see Kerr, 1974). These types of phenomena are generally discussed under catastrophe theory (See e.g., Saunders, 1980)

stocks of an annual species. The life history mechanisms shown by annual species (e.g., squid, penaeid shrimp) also tend to confirm that the environment that most species inhabit is unstable: compensatory mechanisms here include production of many eggs, several separate groups spawning at different times of the year, and a fairly wide geographical range for single stock species (e.g., the Pacific squid, <u>Todarodes</u> which is found throughout much of the North Pacific), thus ensuring repopulation of a stock from the centre of its range if the peripheral populations are wiped out by an unsuccessful spawning.

The existence of multiple age groups is itself then, with multiple spawning, a life history adaption that reflects the difficulty of ensuring species survival in an unstable physical environment, as evinced by perturbations or irregularities in annual recruitment. Recent studies in the North Atlantic on long-term recruitment trends in cod and haddock, show that the very occasional big year classes play a major part here in supporting the fishery in most years; the rest of the time the current level of annual yield would not be sustainable from current production by recruitment, i.e., heavy reliance is placed for a number of years in a row on fishing survivors from previous good year classes. For these North-West Atlantic species, annual landings exceed annual recruitment about 70-80% of the time; from only 10% of the bigger year classes came 24%, 33% and 37% of the total yield for cod, mackerel and haddock, respectively (see e.g. Hennemuth and Autges, 1982). In such a fishery, it seems likely that a series of investment waves also occured, following closely behind good year classes in these fisheries. This is likely to prove typical of what happens in other industrial fisheries elsewhere (Caddy, 1984). Some of the man-induced changes in a fishery that can result from periodic waves of investment, are illustrated in Figure 20. These periodic fisheries investments can be initiated by a particularly good year's recruitment to the fishery, and should be taken into account in socio-economic studies of the fisheries sector. Further approaches to classification of fisheries that take the above points into account, are given in Caddy and Gulland (1983).

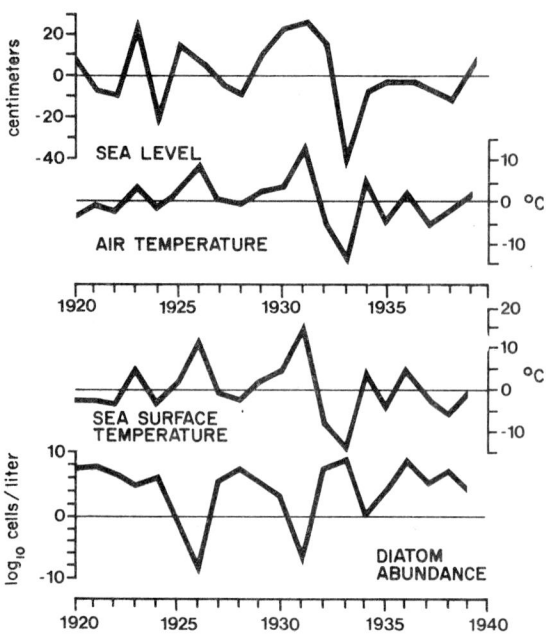

Figure 19 Long-term yearly anomalies in sea level, air temperature, sea surface temperature and diatom abundance at Scripps pier (Redrawn from Horne and Platt, 1984)

2. SYSTEM STABILITY AND MANAGEMENT

The environment: It goes without saying that fish populations, their abundance and species composition, are very responsive to changes in key environmental variables that are not under control of the fisheries administrator. In many cases these variables may act on the stocks in a way that is poorly understood, despite a large literature on effects of a range of environmental variables on growth, survival and reproductive success. Convincing evidence of the impact of environmental change, even without fisheries, has been provided by recent studies of, for example, the occurrence of fish otoliths in stratified bottom sediments from productive upwelling zones; these showed wide fluctuations in abundance and frequent switches in dominant species at specific sites over hundreds and thousands of years prior to human exploitation (e.g., Devries and Pearcy, 1982).

The simplest mathematical representation of a two species system that has been put forward to explain regular fluctuations in linked predator and prey populations, is that referred to as the Lotka-Volterra model and this has been one basis for the modelling of oscillating resources (e.g., Larkin, 1963). it seems likely however, that the extension of such an approach to more than two species will result in a proliferation of population parameters at a greater rate than it will be possible to collect the necessary data to estimate them. This model nonetheless indicates in simple form that an assumption of 'steady state' may have to allow for a degree of oscillation in the abundance of components of a (simple) food web around the mean, even without extrinsic influences. Such oscillations in recruitment - hence population size, in otherwise stable populations, can infact be shown to be initiated by human responses to changing abundance (Caddy, 1984; Allen and McGlade, 1986).

It has also been observed that the behaviour of dominant and subordinate species in an ecosystem may be modified in relation to environment, if they change places in relative abundance in the system. Skud (1982) noted for example, that if all species when dominant respond favourably to increases in temperature, they may show the opposite reaction when subordinate, since they are now suffering population pressures from the dominant species which is more successful with its higher biomass in taking advantage of the changed environment. The point of this observation is that environmental effects, if not understood or taken into account, will in some marine systems make for considerable errors in the yield forecast from a given expenditure of effort, or alternatively, will mean that the effect of a given management measure will be very variable.

Analysing the relative impacts of system instability and intensive fishing is a problem area that has greatly concerned fisheries workers in recent years (e.g., Sharp, 1980b; Sharp and Csirke, 1983), and there is growing evidence that changes in species dominance may be accelerated by fishing. An example is the apparent partial replacement of a sparid dominated community in the 1920s by a community dominated by cephalopod species (Caddy, 1983). Such processes may best be modelled in a discontinuous fashion, and the theory of catastrophes has been proposed as one way of

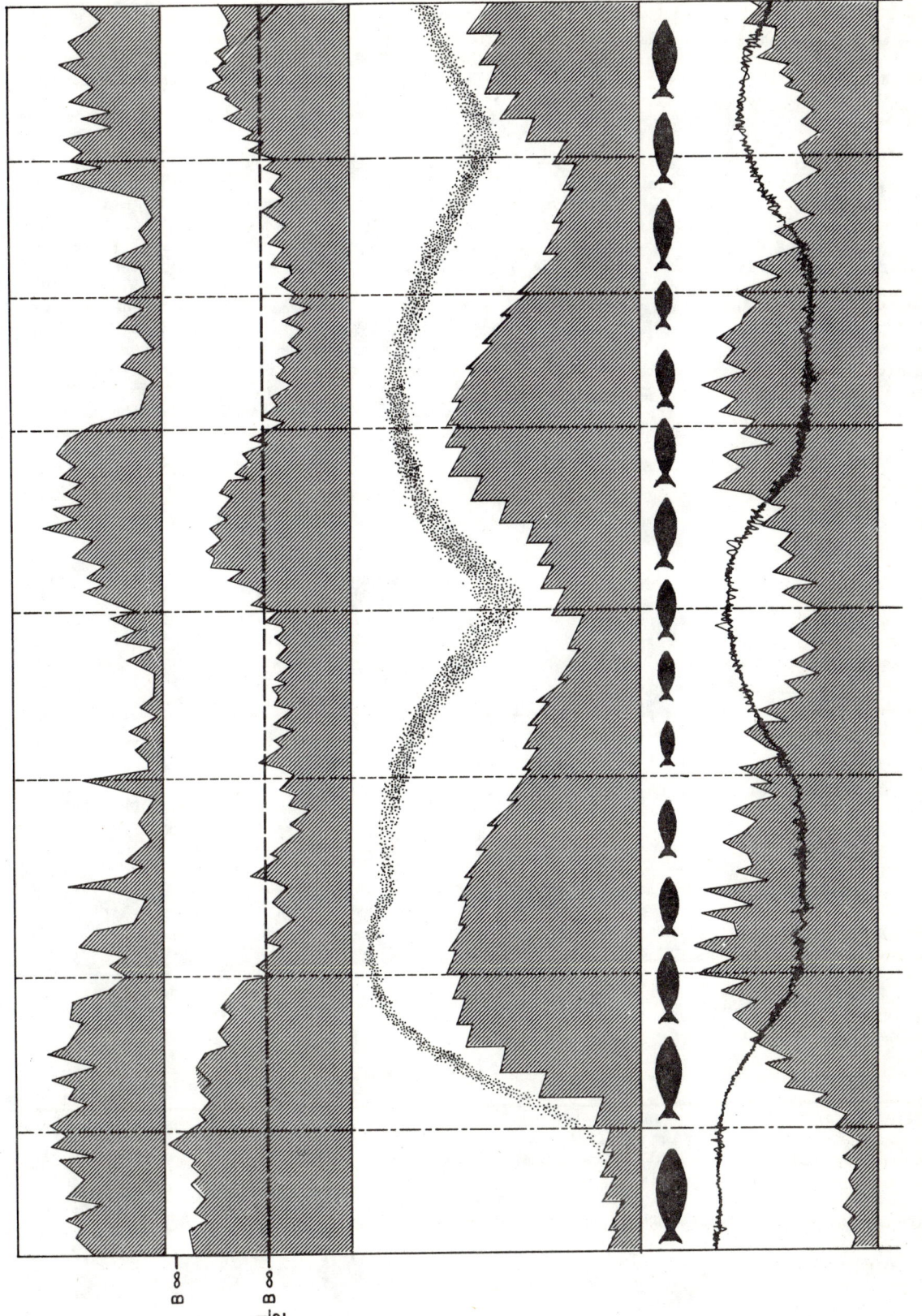

Figure 20 An idealized fisheries "cycle" showing trajectories of some important variables (From Caddy, 1984)

describing this abrupt switching of community dominance (see Figure 18 for an example of a "cusp catastrophe" and Saunders (1980) for some simple theory.

The general conclusion that has been drawn from the phenomenon of temporal variations in marine environments and fish ecosystems, is that there is already a substantial degree of 'elasticity' built into most marine systems during their evolution, and that it is this elasticity that permits harvesting by man on a continuing basis as long as the stresses imposed are not excessive, or do not drive the system into a different state.

This concept of system "elasticity" or resilience does not necessarily mean that ecosystems return directly to their former state after the disturbance. Odum (1969) recognized for terrestrial communisties that the diversity of a commmunity increases in time along a gradient of successional stages as a new (or depleted) habitat in recolonized. It is not clear to what extent this concept of "successional stages" or increasing degrees of community maturity, applies to the marine ecosystem however; given that some highly productive areas (e.g., upwelling systems), seem to be "arrested" at an early, simple stage of community development. However, the progressive colonisation of marine surfaces (marine "fouling") shows some similarities, with a succession of organisms predominating. As Odum notes, estuaries and intertidal zones are maintained in an early, relatively fertile stage by the tides, which provide for rapid nutrient recycling. In complex food webs on land, however, the bulk of biological energy follows detritus pathways, a situation that has some parallels in marine ecosystems described later in this document. Clearly tapping the detritus food web (e.g., by shellfish culture) can lead to some very high levels of marine production. Odum (1969) contrasts the main features of early and late successional stages of ecosystems in the following manner:

Young	Mature
High production	Protection
Growth	Stability
Quantity	Quality

With some qualifications, this rather oversimplified generalization applies in the marine environment also, although as noted by Holling (1973), the more obvious antithesis to stability in the above scheme is resilience, which is not notably a feature of most mature ecosystems, but should be in Holling's opinion, a characteristic of schemes for ecosystem management. In managing a mixture of successional stages, we should, in his words, be prepared to "view events in a regional rather than a local context", and "emphasize heterogeneity".

3. SPATIAL CONSIDERATIONS: MAPPING FISHERIES RESOURCES

The importance at an early stage of a fisheries resource investigation of mapping the spatial distribution of key components in the biological system to be studied is emphasized by most ecologists, but has perhaps been neglected in fisheries investigations, perhaps because of the difficulties of actually determining the distribution of ichthyofauna, and of related environmental and substrate (e.g., sediment type) characteristics, below the intertidal zone. A renewed interest in this procedure has been evident in recent years (Butler et al., in preparation), and in tropical clear water areas, mapping of key features of the shallow sublittoral by remote sensing (satellite) technology has made this more feasible than formerly. From the perspective of analyzing multispecies fishery systems, it is likely that a better understanding of community structure and its response to fishing for demersal fish assemblages, would be one consequence of more precise spatial information on fish distribution and catches, as well as on the areas fished seasonally by the commercial fleet (Caddy and Garcia, in press).

One of the important components of resource mapping is thus the characterization of the principal marine communities or ecological complexes, and their geographical extent. This may be done on a very broad scale initially - e.g., Garcia (1982) for West African exosystems: (see Figure 21 for an example of a mapping of the distributions, migration and stock units for one important pelagic resource); and Baisre (1985) for some generalizations on the distribution of Cuban shelf ecosystems. Baisre divided Cuban fishing grounds into three main complexes, and since most Cuban fisheries are fully or close to fully exploited, fish productivity estimates for each were then possible from separate estimates of their spatial extent. The three complexes are as follows:

(a) The estuarine littoral, including coastal lagoons, estuaries and bays where land effects are important, and where food chains are characteristically short due to high, irregular environmental stress caused by intermittent (seasonal) inflows of fresh water, nutrients and sediments. These areas have the highest productivity; estimated in Cuba at $1.47 t/km^2$ of shelf, they include such coastal ecosystems as mangroves, discussed later, as well as supporting high value fisheries, especially for penaeid shrimp. In Cuba, this area is estimated at $8\ 500\ km^2$.

(b) The second ecological complex in Cuban waters is associated with coral reefs. Coral reef species are of considerable importance, but many of the important fish species here also forage at night on adjacent turtle grass beds, which Baisre (1985) includes in the same complex. This complex is typified by long food chains due to high environmental predictability, and an overall productivity estimated at 0.58 t/km^2, and 45 000 km^2 in extent: spiny lobsters are the most valuable species.

(c) Cuban territorial waters overlap an area of much greater extent, namely the fisheries complex characteristic of oceanic waters, with a fisheries yield of some 0.24 t/km^2 tuna species (especially skipjack and blackfin tuna). To an undetermined extent, some of this production here results from outflows from the preceding two zones.

Trophic diagrams showing the interchange in Caribbean waters between these major ecological zones are given in Figure 48.

Portraying Events in Time

Two types of approaches to the portrayal of the time sequence of events in ecosystem analysis might be mentioned here, which can be graphically represented as an aid to research decisions. The first of these, path analysis, represents an attempt to express the functional and temporal relationships between abundance of food web components, and the physical and biological factors that influence them. Such an approach (shown in Figure 22) for investigations on the causes of stock fluctuations in flying fish), is a prerequisite for time series analysis, and has been formalized under the name "path analysis" This is a procedure for explicitly examining hypotheses, and developing causal models linking time series of correlated variables. A good example of this approach is given in Coelho and Rosenberg (1984).

Flow Charting or Scheduling Research Activities

Recognizing that it is important to represent events sequentially in time, this approach to representing points in a complex train of events is widely used, and may be represented in the case of food web analysis by Figure 8. Such an approach also has obvious applications for planning research programmes (see e.g. Welcomme and Henderson, 1976) and Figures 22 and 23.

Monitoring seasonal changes

Information on seasonal distribution and relative abundance will be needed to distinguish key migratory species from the resident species making up the main fish assemblages (e.g., Tyler, 1971), and in this case, separate distribution maps may need to be drawn for each main season of the year. This information should be supplemented where available, by a comparison of annual time series of abundance or landings of each species. Here, the records of local fish companies and markets, and of buyers and fishermen, may provide useful clues to past and current variability of species catches. Temporal coincidence (common or contrary) and changes in relative species abundance, may be useful clues to possible trophic interactions or competition, if supported by other biological information.

Six types of questions can usefully be asked in the course of a fisheries mapping exercise:

(1) What is the appropriate spatial scale of the investigation?

(2) What is the spatial distribution and relative abundance of those species which are now economically important to fishermen, on a seasonal basis, and throughout the year?

(3) What information is there (in records of fish buyers or the memories of other fishermen) on the extent of past year-to-year variations of the important species?

(4) Which species occur locally year round, and which are transient, occasional or migratory?

(5) What information is there on the diet, behaviour and ecological interrelationships, of the main species in this area or in similar (adjacent) areas?

(6) Are there any obvious seasonal changes in food preferences of the main species of commercial interest? (see e.g., figure 24 for an example of seasonally varying food linkages).

A degree of spatial coincidence between those species that share a common habitat for at least part of the year would thus seem to be a necessary precondition for an important species interaction to take place. Records of the areas of distribution of important species can be provisionally mapped on a seasonal basis from e.g. earlier resource surveys or fishermen interviews. Bearing in mind that this latter kind of information is likely to be less than completely

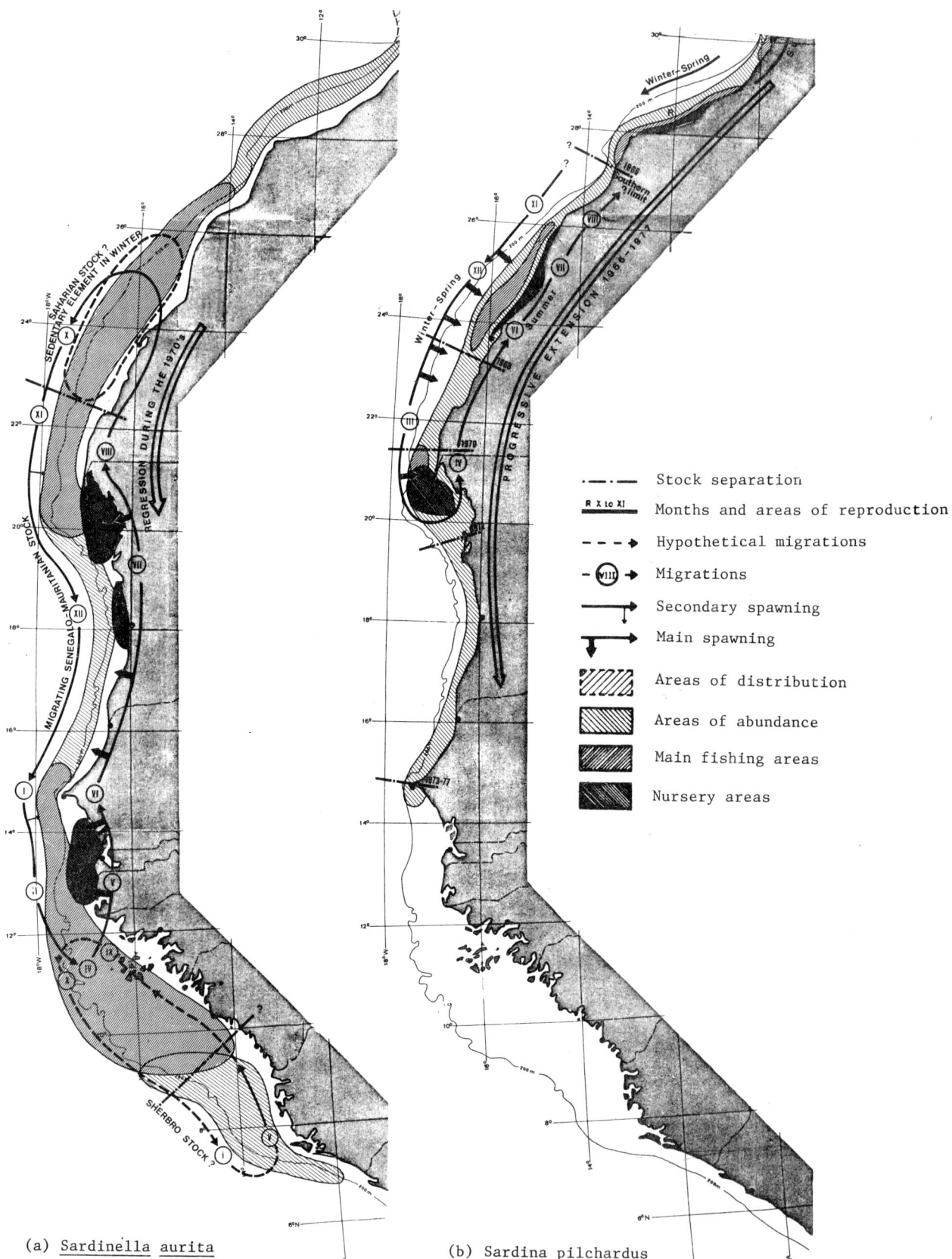

Figure 21 Mapping of seasonal and long-term changes in migration and distribution of two pelagic species associated with the West African upwelling system (From Garcia, 1982)

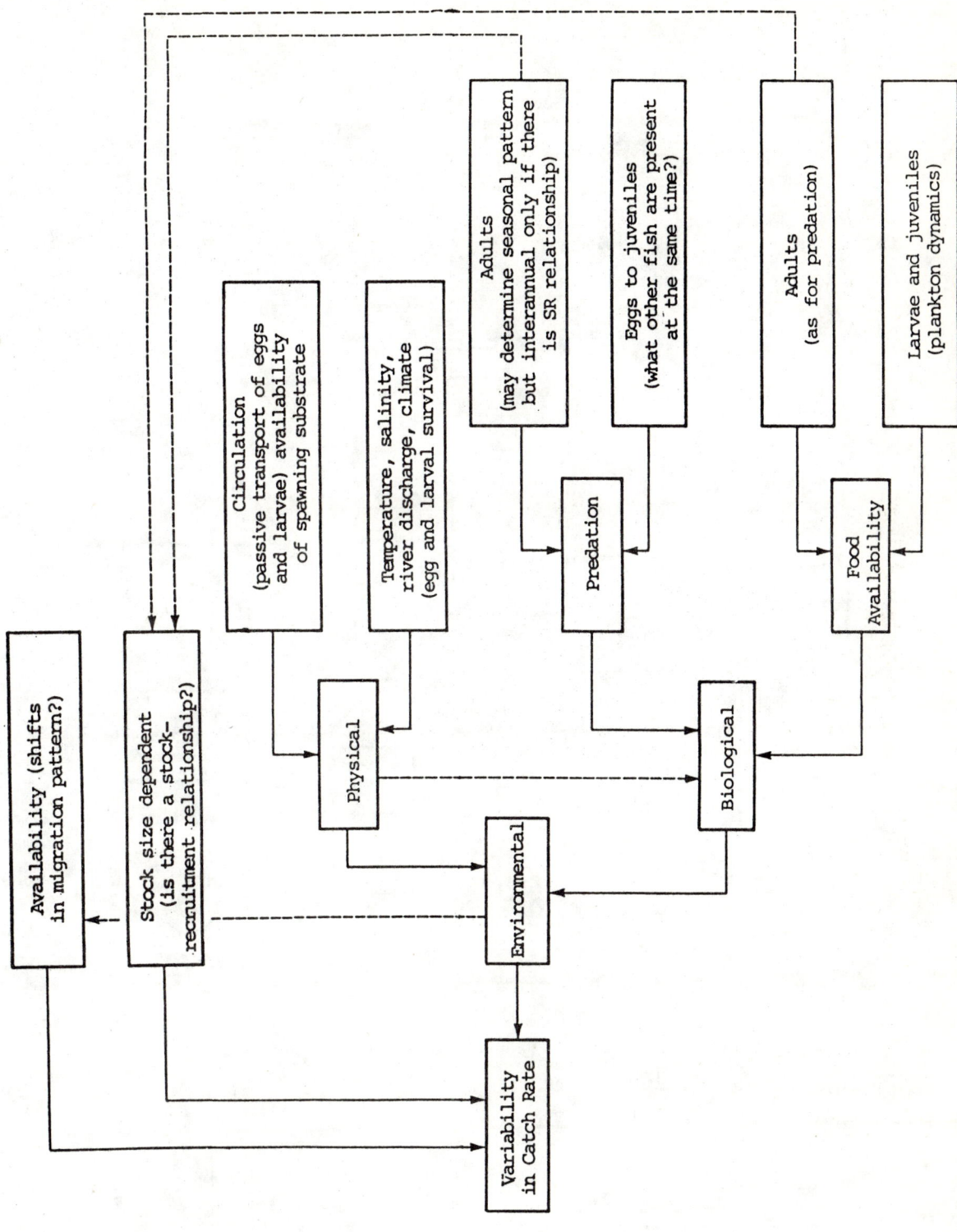

Figure 22 Suite of factors to be considered in evaluating interannual variability of flying fish (From Mahon, Oxenford and Hunte, 1986)

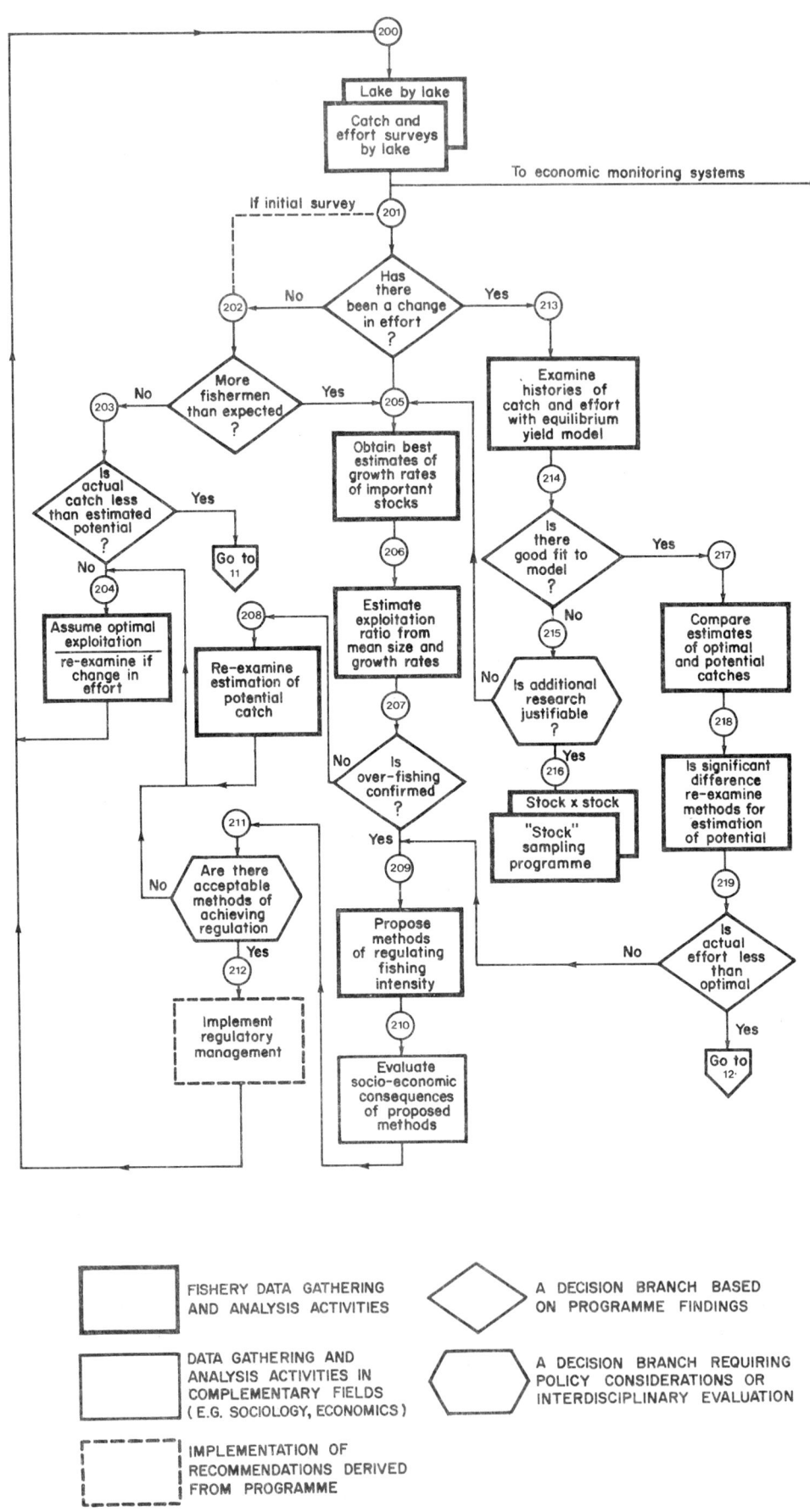

Figure 23 Flow chart of decision-making in management of fisheries
(From Welcomme and Henderson, 1976)

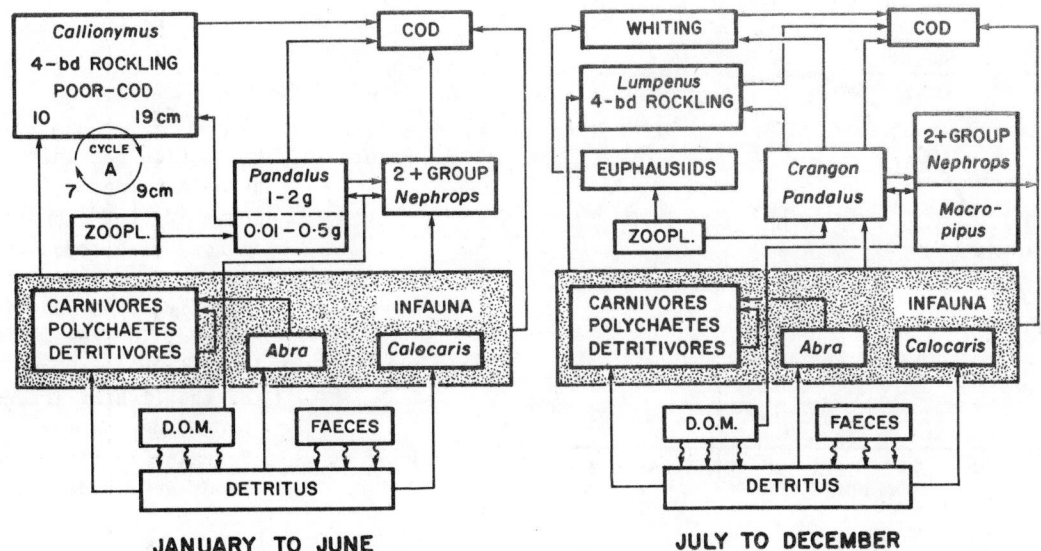

D.O.M.: Dead organic matter

Figure 24   Simplified representation of the feeding relationships of a cod stock showing differences in principal prey at two times of the year (January to June and July to December) (Redrawn from Armstrong, 1982)

objective, it may still be worthwhile to try to assemble it, and compare distribution ranges for key species especially at an early stage in investigation.

4.   FISH PRODUCTION PER UNIT AREA ESTIMATES

The final major group of methods that avoids to some extent the questions of biological detail dealt with elsewhere in this work, is the use of past experience in exploiting ecosystems to suggest what is the expected fisheries productivity of the area in terms of yield per hectare or square kilometre. A summary of some typical levels of productivity is given in Marten and Polovina (1982) and substantiates the impression given elsewhere, namely that productivity falls off along the gradient from estuary, coastal lagoon, or upwelling area, to high seas; although a rather wide variance of these estimates around the mean for each habitat type is typical. Thus, for the shelves of many Caribbean islands, the annual harvest varies from 0.4-0.5 t/km^2 to as high as 18 t/km^2 for specific habitats in the Indo-Pacific (or even potentially 35 t/km^2 for American Samoa - Munro, 1984). This wide range undoubtedly reflects regional differences, but contains potential biases from at least two sources:

(a) Multispecies resources of a similar basic productivity may differ in yield depending on the state of exploitation. This should be taken into account by comparing yield/area with fishing intensity (effort/area) over the same area: (Munro, 1977; Caddy and Garcia, 1983). The yields expressed at some standard fishing intensity may then be more comparable.

(b) Because of imperfect habitat mapping, the extent (area) of productive habitats is normally not correctly measured, and is likely to a greater or lesser extent, to include (less) productive areas in the hinterland.

Both of these sources of error can be partially avoided by specific attention to mapping of habitats and/or ecosystem or assemblage distributions, collecting where possible, data on production by ecosystem areas, or by Assemblage Production Units, APUs, (Tyler, Gabriel and Overholtz, 1982), and also, by considering such ecological units as the 'building blocks' of the fishery management system for shelf fisheries. This approach leads to a consideration of production per unit area as a function of harvesting rate (Figure 25).

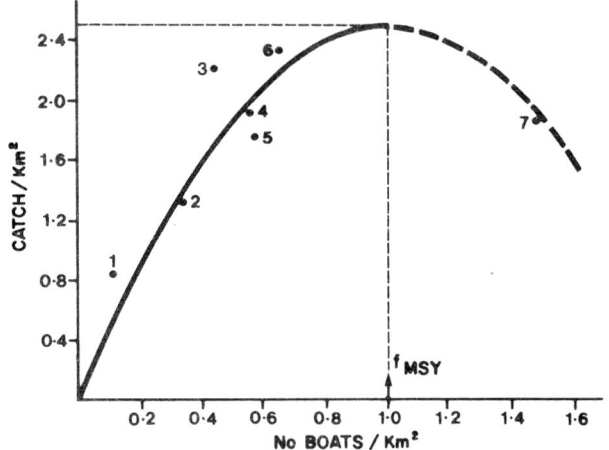

Figure 25 Production per unit area as a function of fishing intensity for a coastal fishery with seven main ports
(Modified from Caddy and Garcia, in press)
(f_{msy} is a rough estimate of the fishing intensity providing Maximum Sustainable Yield under "average" conditions)

5. COLLECTING AND ANALYSING DATA ON FEEDING PREFERENCES OF COMMERCIAL FISH SPECIES

The ability to carry out any serious resource investigation needs, as a first precondition, familiarity with the main taxonomic groups present in an area, and their local common names: (bearing in mind that these latter may vary from area to area). The FAO fish identification sheets offer the most direct approach to identification of fish species and are available, or will soon be available, for many tropical areas (Table 1). These sheets can be used over a period of time to build up a fauna list from species identifications in a given area. If separate files are maintained for each key commercial species, to these can be added direct observations on occurrence by area, season and depth, charts of the main fishing and spawning areas, and other relevant literature. From these source materials, a more or less comprehensive picture of the state of knowledge (and gaps in knowledge) of the species in question can eventually be assembled. The importance of such a "mapping and filing system" for resources is emphasized in Caddy and Garcia (in press) and Butler et al. (in press); and is also a logical basis for development of a statistics monitoring system (Caddy and Bazigos, 1985).

Table 1

FAO Species Identification Sheets for Fishery Purposes (available and in preparation)[a]

Western Indian Ocean (Fishing Area 51)
Eastern Indian Ocean/Western Central Pacific (Fishing Areas 57/71)
Mediterranean and Black Sea (Fishing Area 37)-(New edition in prep.
Southern Ocean (Fishing Areas 48, 58, 88)
Eastern Central Atlantic (Fishing Areas 34, 47 in part)
Western Central Atlantic (Fishing Area 31)

---

a/ Note: A variety of species Synopses and Catalogues summarizing biological information by taxonomic group are also available on request from FAO

Elucidation of trophic relationships

Four main approaches to elucidating of trophic relationships exist, namely:

1. Direct observation methods.
2. Experimental studies.
3. Examination of morphological adaptations of component species.
4. Stomach content analysis.

The first of these methods is particularly feasible in clear water areas (e.g., coral reefs) where it is of growing importance, (e.g., Hobson, 1974). The second class of methods ranges from field experiments involving removal of predators from closed and natural systems and observations on subsequent perturbations (e.g. Paine, 1969), to experiments in field or laboratory conditions on choice of foods. The laboratory approaches will not be described here, involving as they do a significant investment in laboratory facilities and extensive time to achieve results. They are also to an unknown extent problematical, particularly in relation to food preference, which is often difficult to extrapolate to nature, given that the relative availability and abundance of prey is not representative of natural conditions (see however Chapter 12 Part II for an attempt to use existing data on quantities consumed in feeding studies). Despite the significant investment in time and effort necessary for this kind of experimental studies in the field, experiments on caged areas of a bottom community, excluding predators, (e.g., Hancock and Urquhart, 1965; Young, Buzas and Young, 1976), look like being useful in estimating the contribution made to natural mortalities by the various predators, and their role in the food web (Figure 26).

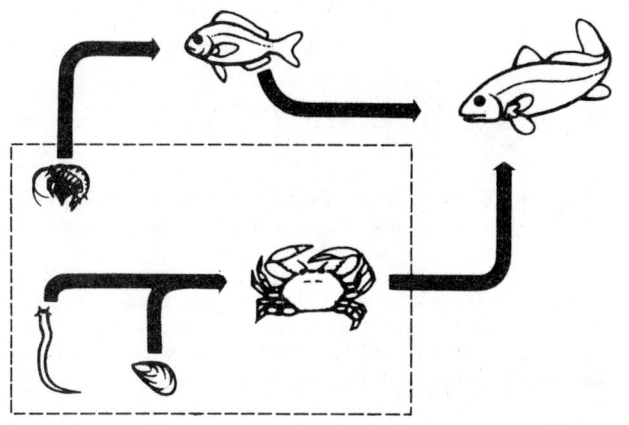

Figure 26   Illustrating an approach to investigating food webs by means of cage experiments: comparison of faunal abundance inside and outside the cage may help to differentiate the impact, for example, of crabs and predatory fish or benthic invertebrates (From Young Buzas and Young, 1976)

The third approach is one which intuitively allows us to distinguish, at least to a first approximation, the main feeding habits of the particular size range of the species in question. This approach relies on an examination of the body configuration and size, and the presence of various morphological adaptations in the species; in particular shape and size of mouth and dentition, form of the gill rakers and intestinal tract, in order to allow a preliminary classification to be made into (for example) planktivores, bottom feeders, grazers or browsers, piscivores, etc., although this type of approach rarely substitutes for method 4 in determining trophic interrelationships.

Stomach content analysis will of course present more complicated problems of identification because of the small size and poor preservation of the stomach contents than direct sampling of the prey, but may provide a more extensive species list and size ranges than that obtained from the commercial catch. This is because a number of species and size will be encountered, possibly for the first time, that are not commercially important or catchable with fishing gear. The problem of species identification can be approached initially by identifying unfamiliar species first to genus or family, and, secondly, in a first (and probably qualitative) approach to food web analysis, only attempting more specific identifications on well preserved and documented specimens after consultation with (museum) taxonomists who are experts in the various groups. In the meantime, an empirical key must be developed for unknown but readily characterized taxa that will at least allow the work to continue (see e.g., Wildish and Phillips, 1974).

Stomach content analysis, even in its most casual and anecdotal form, can yield incidental but immediately valuable information, since predators are often better sampling devices than most commercial fishing gears. The presence of a number of valuable resources (e.g., squid, crustaceans) not yet commercially harvestable from existing gear, have been first identified in an area from this type of observation.

Although entire food webs have been elucidated based on gut-content analyses (e.g., Roger and Grandperrin, 1976), caution should be exerted against over-enthusiasm for large-scale systematic studies of this kind which can be particularly expensive in using up scarce and valuable manpower, research vessels, laboratory workspace and storage facilities, as well as valuable computer time. This appears to be one of those types of activities that very soon leads to rapidly diminishing returns with time. A great deal can however be achieved over a period of time if this type of observation is fitted in with other field work, and we may note that such an approach is a necessary precondition to an understanding of species interactions. Observations on freshly-caught individuals of known species, sex and size can be accumulated over several years when opportunities permit, and the information stored on file cards or in folders marked by species (predator or prey) names, allowing a picture of the trophic interrelationships to build up gradually, which can eventually be abstracted and described. Each individual record should note, in addition to the species of prey and their size and numbers present, the place, date and time of capture, depth and other observations of possible interest; cross-referenced to other sources or records of information collected at the time - e.g., catch rate and size composition of the commercial catch, depth, bottom type as well as fishing gear used, as well as to any museum or otolith specimens that may have been retained.

A description of the mechanics of analysis of stomach contents is given by Bagenal (1978). Four major problems should be recognized before collecting data so that precautions in sampling can be appropriately taken:

1. use collection methods that minimize regurgitation of food (such as, for example, caused by gillnets);

2. avoid holding fish for protracted periods (e.g., long lines, traps) before removing and histologically fixing stomach contents;

3. given that most fish species have a relatively restricted period of feeding, timing of capture should ideally take into account the diurnal feeding cycle and also seasonal variations in diet, where these occur;

4. the rate of digestion (or more relevant, the rate of loss of identity of the food item) is more rapid for soft-bodied prey than for those with pronounced exo- and endoskeletons, thus making greater problems in identification, and exaggerating the contribution of the latter type of prey species to the diet if these are retained longer in the gut.

Gravimetric, or if not possible, volumetric evaluation of the gut contents in order of exactness, is the recommended approach in most texts. We may note, however, that for a preliminary investigation of the type we are suggesting here which is not principally concerned with the energetics of biomass transfer, enumeration and measurement or estimation of length ranges of fish prey species is largely adequate, and may be of more importance than total species weights in terms of recognizing critical stages in predator life histories. The first thing that may be evident and of potentially commercial importance, is to recognize that certain changes in feeding preferences occur seasonally and in the life history, and to note the periods and sizes at which these occur. Any study of the species may then recognize these size categories in field sampling. A more direct application is to recognize that in certain areas and seasons, a species may be available to capture in a fairly limited zone (or season) where a particularly characteristic and locally available food species occurs. Information on such (potentially commercial) prey species may not always be evident from experimental fishing or survey results.

Preliminary food web analysis

As we have noted, studies of trophic interactions in northern latitudes are based upon a long and rich scientific literature on life history, food preference and behaviour for a system where relative abundance and size and age structure of most fish populations are rather well known. Under these conditions, the construction of complex interspecies models (e.g., Anderson and Ursin, 1977; Laevastu and Larkins, 1981) is a logical extension of a rich data base, as is the extension of other conventional methods of single species analysis to multi-species use (e.g., phalanx analysis; Pope, 1980). These rather detailed approaches to multispecies assessment are less likely to have any early application in data-poor situations, however, (and we may note that these models are not usually applicable on a routine basis, even in better-studied areas). Larger models with many parameters, in fact pose serious problems to the fishery scientist of a philosophical nature: if all the information is included in the model, what data should be used for testing it? Hence we may have to be content with some understanding of the structure and main interactions of the system, and in consequence, expect to gain at best, some qualitative guidelines for multispecies management.

At first approximation, it may be convenient to proceed directly from a preliminary prey-predator contingency table (e.g., Figure 42) to construction of a tentative food web (Figure 43), even though estimation of the rate of feeding of predators in absolute terms requires a considerable knowledge of the ingestion, digestion or egestion rates, both as a function of environmental factors and of type and abundance of food. These latter factors will be touched on later, but for the moment the assumption is made that the relative volume of identifiable food contents per predator body weight is an index of their fraction in the total diet. This assumes that the sample of stomachs is properly weighted both for diurnal and seasonal factors, and for the distribution and relative abundance of the different size groups under investigation.

To some extent, a preliminary trophic classification of this kind is possible by reference to the literature, although information is surprisingly sparse on the diets of many marine fish species.

"Stomach content analysis" (including intestinal contents) is the principal method available to the field worker with limited facilities for intensive laboratory or field studies. This methodology, and the data it generates, has been inadequately utilized to date, mainly, we believe, because of uncertainties in the interpretation of the large volume of data that can be rapidly obtained by this method. To a significant extent this is also because methods of population analysis until recently have placed inadequate emphasis on multispecies approaches and species interactions.

6. EQUILIBRIUM CONCEPTS AND THE FLOW OF ENERGY THROUGH A SYSTEM AND ITS OPTIMAL UTILIZATION

6.1 The equilibrium concept in fisheries

It may be useful when discussing ecological stability, to follow Johnson (1981) in contrasting the concept of an isolated system coming to a unique thermodynamic equilibrium, with the term equilibrium as it is normally used in the fisheries literature.

In thermodynamics, the flows of energy and/or materials at equilibrium across a boundary between zones A and B: i.e., $A \rightleftharpoons B$, are equal and opposite; implying a state of rest exists, where there is no net transfer of energy and materials.

In population dynamics, the rate of flow X of energy from the rest of the food chain, to the component of population biomass (B) being subject to fishing, is shown in the following simplified scheme:

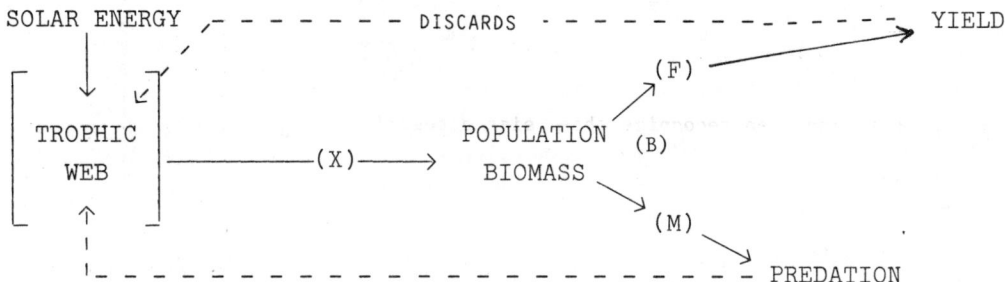

The rate of flow of chemical energy in the form of biomass tapped by man, and by natural predators are respectively the (ponderal) rates of fishing (F) and natural (M) mortality, as discussed in the literature on fisheries assessment. The flow of materials of immediate concern to the fishery is from left to right in the above scheme and although, eventually of course, there is a reverse flow as shown by the dashed lines, this does not constitute the "equilibrium" referred to here, which is said to occur when a constant yield can be removed from the population biomass per time interval, without prejudicing the systems' ability to sustain this yield. The yield can then be considered "surplus" in that it does not affect the stability of the population. Of prime interest to fisheries ecologists then, are those factors affecting the rate of flow (X) of energy and materials from the food web into the exploited system, and how variations in the energy entering the system, and the structure of the food web itself, affect the yield from the system. Clearly this yield cannot be considered as being strictly at equilibrium with biological production unless both production and harvesting rate are constant or unless the latter is closely responsive to the former. Any "equilibrium" that may come to exist, will therefore not be passive, but will depend on a good system of monitoring of both production and fishing effort, and on a constant M.

Before looking at an ecological system from the point of view of its energetics, we could begin with an analogy, by examining a better known system which is currently under intensive discussion, namely the different uses of energy by modern industrial societies.

Figure 27 shows in a diagrammatic form how society makes use of the various potential energy sources available to it, by in most cases transforming raw materials into another form that is most suited to the many and varied applications modern industrial society has found for energy. Several things are noticeable there.

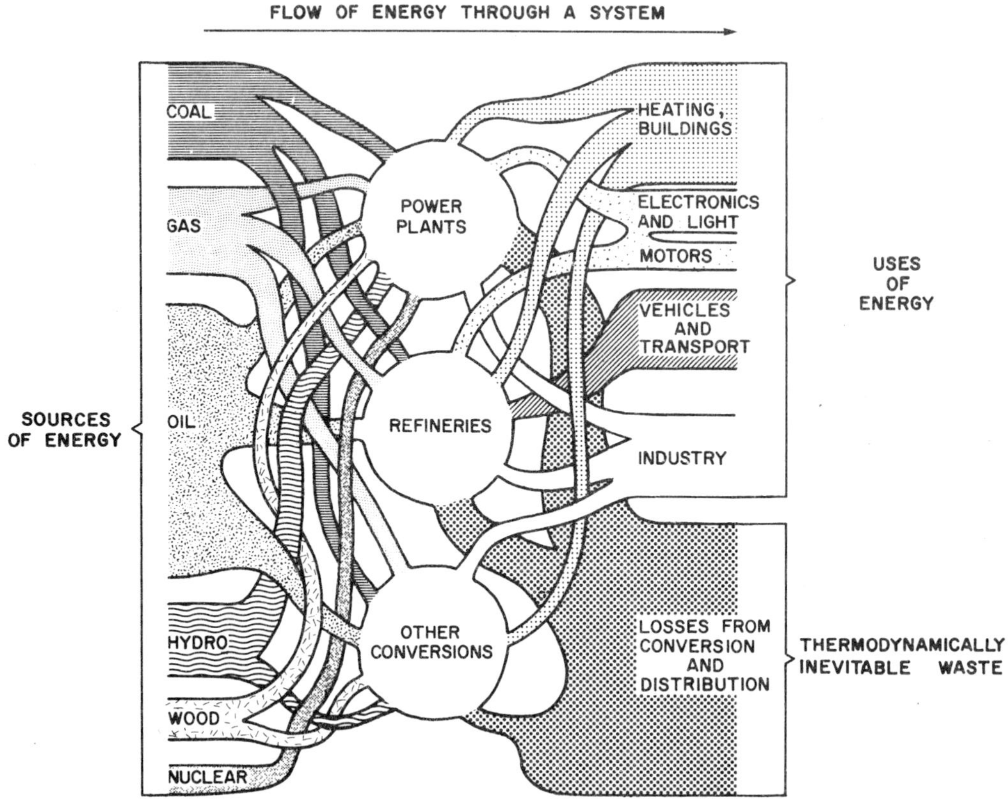

Figure 27   A stylized "spaghetti chart" showing uses of energy in modern society
(After Louins and Louins, 1981)

1.   Figure 27 is in fact an "artificial" dissipation structure in which energy is transformed from a less acceptable form into one that is more easily used by society; noting from the first law of thermodynamics, that although energy can be transformed from one form to another, it cannot be created or destroyed.

2.   A large part of the input energy is "lost" during both conversion and use by man, and (a consequence of the second law of thermodynamics, but not shown here), still more energy must be expended in order to rectify some of the side effects of the use of the energy by man. These result from the need to dispose of the physical by-products resulting from consumption of fuels, (i.e., chemical and physical pollution), and of waste energy: (so-called thermal or heat pollution). These 'down- stream' effects of human interventions also apply in natural systems (see later).

3.   Not all sources of energy are equivalent in "quality" in our modern society − in fact chemical energy (oil derivatives) are "high grade" energy sources and less easily substituted for in their applications than, for example, coal. In other words, in order to use coal for running a car, it is necessary to first transform it physically or chemically, and in doing so, heat (energy) is lost, and a variety of costs incurred; i.e., the energetic efficiency of the overall system is reduced.

   Each of these three features of the above system are equally valid for optimal utilization of a food web − also properly regarded as an energy dissipation structure. This transmits the (still largely unusable) solar energy falling on the ocean into chemical energy, i.e., phytoplankton, by photosynthesis into sugars and other carbohydrates, and subsequently to fats and proteins. This energy is large in quantitative terms, but low grade (i.e., a greater portion is unusable in quality) and/or dispersed in space and time so as to be difficult or costly to harvest (as phytoplankton). Each step upwards in the food web results in a significant energy loss − of the order of 70-90 percent per step − of that initially available. After 1-3 more steps, the food energy may in some forms be considered "higher grade", i.e., more acceptable or desirable as a human food item. Equally important, it may be harvestable in a cost-effective fashion (i.e., occur in larger unit "packages") which, especially in the case of schooling fish, or species that can be aggregated by bait or other attractants, can be harvested yet more cost effectively as a group or school. Obviously there are many exceptions to the idea that level in the food chain is directly related to perceived quality in human terms. For example, in terms of marketability ("quality"), tuna and abalone are higher-priced products on the fish market. Each however, has several different features ecologically:

1. Abalone is a herbivore and therefore near the bottom of the food pyramid. It is also easily harvested as well as locally vulnerable to overfishing, but suffers from a major disadvantage in that its substrate or 'niche', and hence its distribution, abundance and biomass, are very limited.

2. Tunas are highly mobile apical predators in their harvestable stage, but with a habitat or niche that is very large in extent. Obviously only a small fraction of the chemical energy originally synthesised in the oceanic habitat as biomass is available for harvest by man as tuna, and even then, only with considerable investment in technology and effort. On the other hand, the ecological system supporting tuna is very large, diffusely patchy, and the energy needed to harvest and utilize most of the lower trophic levels leading to tuna, would be prohibitive with current technology, fuel and labour costs and market prices. Hence, harvesting a large schooling fish such as tuna becomes an acceptable alternative to harvesting the less available and lower priced, but potentially more productive trophic levels lower - down, (see Chapter 13 on the oceanic tuna system).

Systems concepts of this kind offer us messages that are potentially of great interest to fishery managers, which will be touched upon in greater detail elsewhere in this paper, but can be summarized as follows:

1. In deciding to harvest one or more components of a system, one should explore first the present uses made of the chemical energy by the other components of the unexploited ecosystem and by man. Viewed as a dissipation structure often evolved over millions of years, ecosystems contain an elaborate systems of feed-back loops that means that no interference with the system outputs will go entirely unnoticed by the system, (although a generally high level of system "damping" is the case). Neither will these perturbations necessarily be "harmful", especially if the man-induced perturbations lie within the natural dynamic range of the system's capability to respond to normal variations in environmental or biological variables.

2. When looking at a system before harvesting, we should consider each component from the point of view of (a) its volume (biomass), (b) its availability and vulnerability, and cost of harvesting, (c) its marketability (value) or ease and cost of transformation into alternative more acceptable or desirable products, and last (but far from least), (d) the impact of harvesting any component on other components of the system, or on the system itself.

3. An exploration of how the system components interact will not give you hard and fast answers to all these questions, but will allow you, or a "manager", to identify the risks of any particular action.

4. One of the main uses of a knowledge of trophic interactions will be that 'down-stream' effects of human interventions on the ecosystem may be better visualized in advance (although much more information will be needed before these can be evaluated quantitatively).

Benefits from harvesting an ecosystem

Expressed in simple terms, the benefits from any given strategy of harvesting an ecosystem consisting of $i = 1, 2, 3, \ldots n$ species, is the sum of the benefits from each of the components, i.e.:

$$\begin{bmatrix} \text{Net Benefits } (B_T) \\ \text{from harvesting} \\ \text{ecosystem} \\ \text{(component } i) \end{bmatrix} = \sum_{i}^{n} \begin{bmatrix} \text{Gross earnings} \\ \text{from harvested} \\ \text{component } i \end{bmatrix} - \begin{bmatrix} \text{Costs of harvesting and} \\ \text{processing } i \end{bmatrix} \pm \begin{bmatrix} \text{Other unspecified} \\ \text{gains or losses} \\ \text{due to harvesting } i \end{bmatrix} \quad ..(A)$$

The last term in equation (A), the Interaction term, although difficult to quantify, is essential for generality: it covers those incidental losses, benefits to society or down-stream effects, that will sometimes result from a harvesting strategy, but which are not contained within the fisheries subsystem being considered. Thus, for example, the commercial harvesting of individual components of an ecosystem (e.g., using coral heads for building material in the Indo-Pacific or commercial gillnetting of Atlantic salmon), may each be cost effective when considered as isolated activities, but will also have incidental effects on local revenues from the tourist, sports angling, or sport diving industries. In the first of these two examples, it may also result in deleterious ecological changes, since coral is itself a substrate species supporting reef fisheries, as well as a natural protection against coastal erosion, so that coral harvesting will affect the viability of the system as a whole. Similarly, for a marine trawl fishery to develop, provision of freezer of harbour facilities for the fishing fleet may be a necessary cost for developing the resource, but may also offer other benefits to the food industry or to coastal transportation (see also Chapter 7 on the mangrove ecosystem). This Incidental Component can sometimes (but not automatically) be discounted, but should not be ignored, since it may be considered to represent the linkage between the fishery system, the ecosystem as a whole, and the rest of society, its economy and objectives. Approaches to understanding the economics of natural systems within the constraints imposed by the biosphere, are now beginning to enter the literature (e.g., Passet, 1979).

No theoretical tools yet exist that permit an a priori determination of the optimal level of harvesting of a whole ecosystem; but if broad enough data are collected, we can consider the impact of a given change, assuming that the change in harvesting strategy is sufficiently marked and has been in place long enough for a new more or less steady-state condition to be reached. The impact of a change in harvesting of any component (i) on the marine sector can then be given by the following expression:

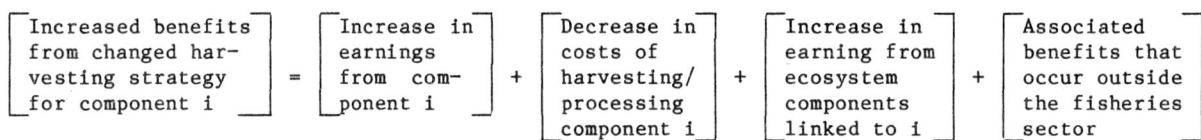

Obviously decreases (-) can be substituted for increases in the above expression with a change of sign on each of the boxes. Implicit in the above equation is the realization that decisions taken with respect to harvesting of one component will have impacts on the other components of the fishery system, and could even have economic impact outside the fisheries sector, as for example, if an increase in coral fish abundance had increased revenues from tourism. Even if we are rarely in a situation when we can define these impacts quantitatively, it is possible under some circumstances to give an idea of the nature of the incidental effects likely to ensue, even if not of their extent.

An example of such a check list of renewable uses and impacts is given in Table 2 for coral reefs.

Table 2

Some direct and indirect potential uses of coral reefs
(modified after Kenchington and Hudson, 1984)

Potentially
Renewable resources uses:

  Harvesting reef-associated organisms

  Controlled fishing:
  - commercial
  - subsistence
  - recreational

  Other: Controlled collecting of:
  - Decorative coral
  - Shells
  - Aquarium fish
  - Turtle, dugong and marine crocodile hunting
  - Bird hunting

- Resort tourism (marine parks)
- Sea-borne tourism (including nature appreciation)
- Air-borne tourism (scenic flights)

Potentially non-renewable uses:

- Limestone and guano mining
- Fishing with destructive methods (e.g., with dynamite, bleach)
- Uncontrolled fishing and harvesting of biota
- Discharge of near shore effluents from industry, tourism and agriculture
- Uncontrolled coastal development

---

7. QUALITATIVE CONSIDERATIONS IN ECOLOGY - THE MANGROVE ECOSYSTEM AND CORAL REEFS

In its most fundamental form, a knowledge of how the various species in our environment are functionally linked together has been part of the human heritage since prehistoric times. The limits on production of the systems they live in are often enshrined in the fundamental religious beliefs of hunter-gatherer societies (and in fact encompass the idea that man is part of nature also). Unfortunately inhabitants of our modern technological society are not automatically aware that the interrelationships which exist between components of an ecosystem are not infinitely elastic, and societies still act in many cases as if these matters pertained to systems extrinsic to their own lives and local micro-environments, whereas it is clear that all environments have a limited capacity to withstand extrinsic stresses.

A good example of the complex and far-reaching impacts of such perturbations is the mangrove ecosystem - a highly important coastal environment common to tidal and brackishwater areas of tropical and subtropical areas of the world (see Tables 3 and 4). Figure 28 from "Conservation Indonesia" gives a simple illustration of some of the typical interdependencies existing within mangrove systems, showing the natural pathways of material from leaf detritus and other plant production inside the mangrove forest, and its subsequent dissipation outwards and upwards in the food chain. This material enters the aquatic system as leaves and sticks, and later in the form of detritus and the excreta of herbivores, and is used as a substrate for an abundant bacterial flora. It then passes in sequence through the bodies of detritus feeders (shellfish and other invertebrates), primary (small) carnivores, and secondary and higher (large) carnivores. Also briefly illustrated here are the multiple usages of various components of this food web by the human inhabitants of the coastal ecosystem. Just in terms of natural products harvested by man, the timber itself provides fuel and a diversity of construction materials, including a number of valuable and inexpensive components for the construction of fishing gear; plus a wide variety of foodstuffs, folk medicines, agricultural and other products (Saenger, Hegerl and Davie, 1983); not

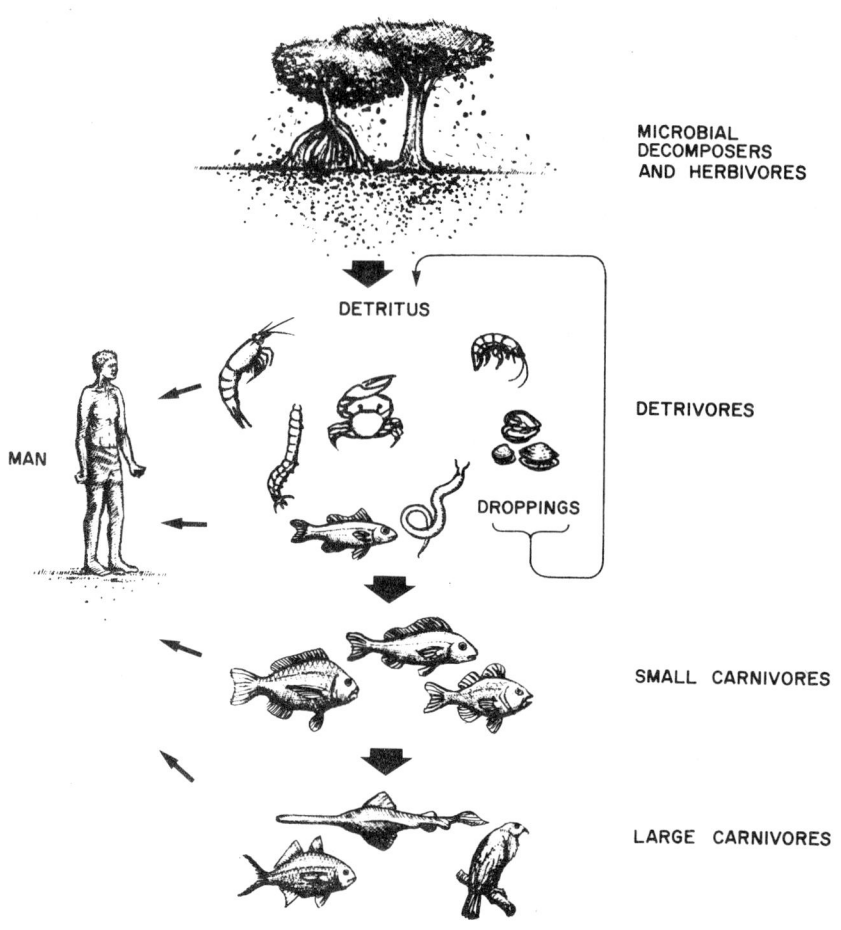

Figure 28 Interdependencies and dissipation outward of biological production in the mangrove environment (Modified from "Conservation Indonesia", Vol. 5, No. 3, December 1981)

to forget the key role of this system in the life cycles of other components of the coastal ecosystem, including man, e.g., Figure 29. Kapetsky (1985) estimates some 460 000 artisanal fishermen live around mangrove associated estuaries and lagoons (Table 4). Beginning first with the production of plant material, Christensen (1978) estimated for a stand of Rhizophora in southern Thailand that annual production of organic matter above the ground was equivalent to 7 tons of leaves and 20 tons of wood per hectare per year. This amounts to 10 tons of plant production that is ultimately not used by man (twigs, leaves, fruit, etc.). This "deciduous" component of production per m^2 of mangrove forest is roughly five times the phytoplankton productivity of the same area of coastal zone without mangroves, and Kapetsky (1985) estimates very high figures for production from mangrove-fringed estuaries and lagoons (91 kg/ha/yr). World-wide, this is roughly 925 000 t of fish and crustacean products coming from coastal and estuaries areas where mangroves are important components of the ecosystem. This almost certainly implies (as for other coastal macrophyte production systems, e.g., kelp and sea grass beds), a large net export of detrital material to the food chains of the continental shelf. Mann (1972), for example, estimates that roughly 40 percent of the detrital material produced by macroalgae in high latitudes is dispersed offshore to adjacent shelf areas and eventually to groundfish food chains.

The other roles of the mangrove complex in the ecosystem and life history of the associated fauna are well marked: close to 30 species of fish, 200 species of crustaceans, over 200 species of birds, and a wide variety of species from other taxa make this a highly diverse community with a complex food web that is much interlinked with adjacent ecosystems, and would therefore (like other highly diverse environments, such as the tropical coral reef), be impossible to portray as a

Table 3

Some gross economic values of mangroves for fisheries
based on value of landings (from Kapetsky, 1985)

Country/Region	Value in US Dollars	Basis for value measurement	Year of estimate	Principal species
Panama/ Gulf of Panama	26 350 (/km)	Value per kilometre of mangrove shoreline	1978	Penaeus, Trachypenaeus
	65 164 (/km)	Value per kilometre of mangrove shoreline	1978	Cetengraulis, Mysticetus
	3 114 (/km)	Value per kilometre of mangrove shoreline	1978	Micropogon, Lutjanus, Centropomus
Brazil/ Cururuca Estuary	76 886 (/km^2)	Includes only finfishes. Estimate based only on areal extent of open water	1981/82	Mugil, Genyatremus, Macrodon, Bagre, Macropongonias
Malaysia/Sabah	133 (/km^2)	Value based on areal extent of mangroves	1977	Scylla serrata
Malaysia/ Peninsula	277 235 (/km^2)	Areal extent of mangroves plus estuaries and lagoons	1979	Penaeus, Stolephorus, Pampus, Polynemus, Lutjanus
Thailand/ Khlung District	3 090 (/km^2)	Value of fishery products captured inside the mangrove system	1977	Liza, Eleutheronema, Arios, Ophichthus, Lates
	10 090 (/km^2)	Value of mangrove-associated species caught elsewhere	1977	Penaeus
Bangladesh/ Sundarabans	2 076 (/km^2)	Value based on mangrove area plus open-water area	1982/83	Hilsa, Penaeus
Papua New Guinea/ Gulf Province	476 (/km^2)	Value of shrimp caught outside the mangroves, and of subsistence fishing and crabbing inside	1977	Penaeus, Metapenaeus, Scylla serrata, ambassids, gobies, gudgeons, catfishes

detailed food web. Some degree of compartmentalization of the system into herbivores, detritivores, primary predators, etc., is necessary before going much further in describing such a system. However, even here changes in trophic level during the life history are common (Table 5), and make this objective far from easy to attain. The detailed description of such a complex system with many external linkages as a 'closed' food web, may not therefore in itself be very helpful, especially bearing in mind that the coastal, and especially the intertidal, zone is a transition area between terrestrial and aquatic systems, and as such is completely "open-ended". The mangrove system receives inputs from terrestrial systems in the form of runoff of freshwater containing dissolved nutrient salts, organic material and growth factors, bacteria and often, pollutants either in suspension or adhering to silts, clays or detrital material. The system provides similar outputs to offshore areas, as well as providing habitats (nurseries) for many coastal or even offshore species of commercial importance at one or other stages of their life history.

It would be necessary therefore, in modelling such a system trophically, to show the major routes whereby the contributions to, and losses from such an open system occur, if a food web is to be of much predictive value in this case. The economic importance of such 'transition areas' as estuaries and lagoons to fisheries, is documented in various places (see Table 3), and is considerable. For example, it is estimated that as much as 90 percent of the U.S. commercial catch, and 70 percent of the recreational catch in the Gulf of Mexico, is made up of fish, shellfish and crustacean species that spend all or a vital part of their life history in estuarine areas (where mangroves form an important part of the habitat). It has even been demonstrated (Fuller, 1979), that the length of shoreline in estuarine areas is a controlling factor in shrimp production: something to consider when "rationalizing" shorelines by landfill in coastal wetlands.

Table 4

Derivation of the mangrove fishery yield and
mangrove fisherfolk estimates (from Kapetsky, 1985)

A. Total world mangrove surface area = 171 000 km^2 (Rollet, 1984)

B. Total open-water surface area (lagoons and estuaries) associated with mangroves estimated according to:

   Open-water Area = 0.481 (Mangrove Area) + 248.3 (see Figure 3) = 82 535.3 km^2

C. Median yield of finfishes, shrimps, and crabs from 18 mangrove-associated lagoons and estuaries = 9.1 t/km^2

D. Annual yield of finfishes, shrimps and crabs from coastal lagoons and estuaries associated with mangroves = 82 535.3 km^2 x 9.2 t/km^2 = 751 071 t/year

E. Annual yield of molluscs from mangroves based on one example only (Bacon, 1984) equals 2.1 t/km^2, or 175 424 t/year when extrapolated for the total mangrove-associated open-water area (meat weight)

F. Median fisherfolk density in 14 mangrove-associated coastal lagoons and estuaries = 5.6 fisherfolk/km^2 (see Figure 4)

G. Total fisherfolk of coastal lagoons and estuaries associated with mangroves = 82 535.3 km^2 x 5.6 fisherfolk/km^2 = 462 196 fisherfolk

---

Here we may specifically mention the penaeid shrimps, some species of which, although spawned offshore, spend their critical inshore juvenile stages in coastal lagoons and mangrove dominated areas (see Garcia and Le Reste (1981) for a review). Some indications have emerged of an apparent regularity between the declining extent of mangrove areas and the decreased shrimp catch offshore from each coastal area, and a similar relationship between mangrove area and offshore shrimp production from different regions of the coast of the Philippines is suggested in Martosubroto and Ndamin (personal communication).

It seems as if it is necessary to consider all potential usages of these habitats and, to look at the linkages to adjacent offshore environments when considering major changes or alternate uses of these areas for (say) building, landfill disposal, forestry, "wild" fisheries, salt pond and fish pond use, etc. Similarly, on the terrestrial side of the land-se interface, mangroves perform a useful function as windbreaks for tropical hurricanes, help in flood control, and prevent soil erosion. All of these factors and more are discussed in Christensen (1978), and also need to be considered before major ecosystem changes are made: (see cover design).

Table 5

Changes of food habits of fishes from Crystal River, Florida, correlated with ontogenetic growth and dietary group (from De Sylva, 1975)

I.	Planktivore throughout	VI.	Transition from planktivore to cleaner carnivore
II.	Transition from detrivore to planktivore	VII.	Carnivore throughout
III.	Transition from planktivore to herbivore	VIII.	Transition from carnivore to omnivore
IV.	Transition from planktivore to herbivore to carnivore	IX.	Transition from detritivore to omnivore
V.	Transition from planktivore to carnivore	X.	Detritivore throughout

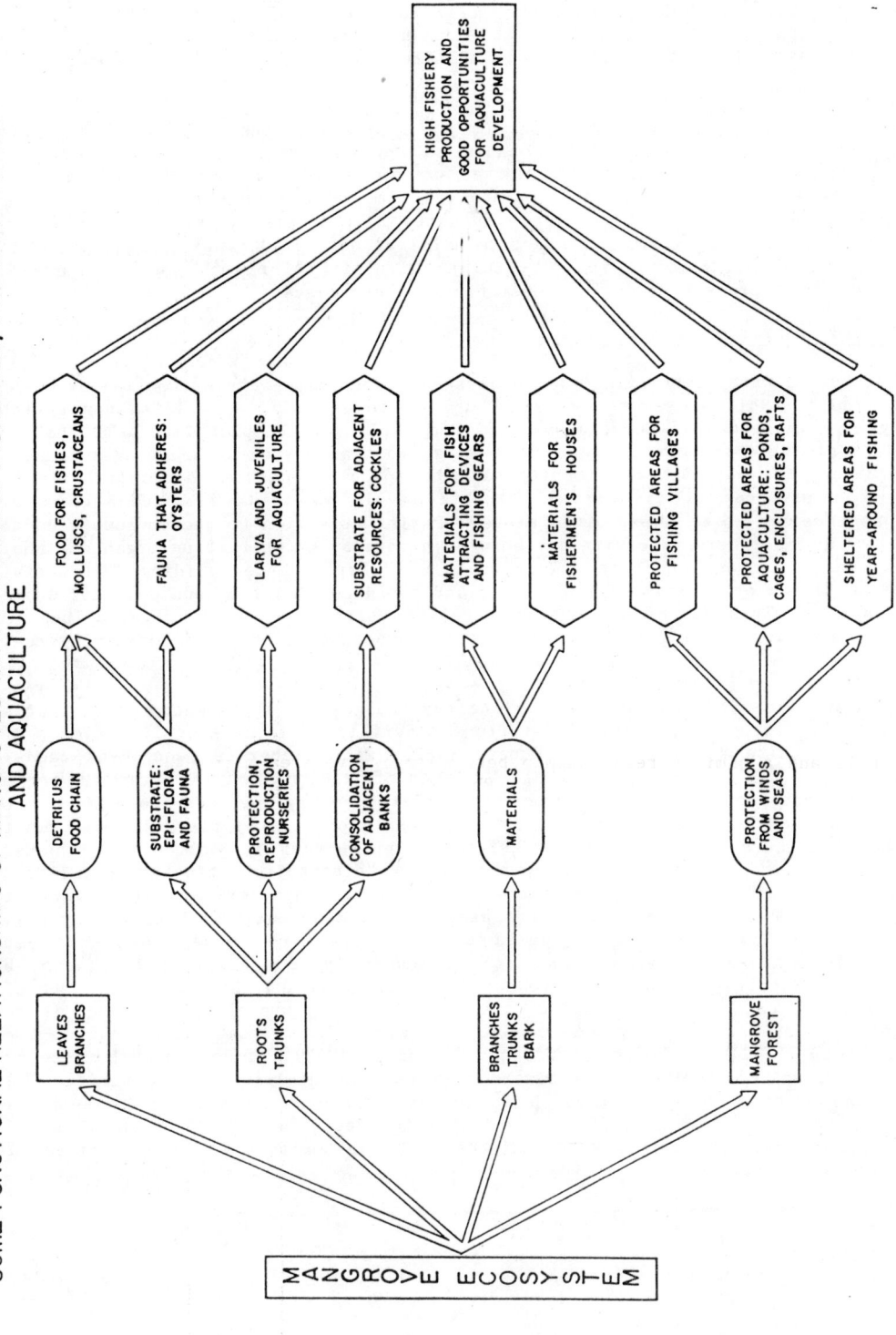

Figure 29 Some functional relationships of mangroves with artisanal fisheries, fishery resources and aquaculture (From Kapetsky, 1985)

In conclusion, although this system is one of the more complex in the world in terms of species diversity and interactions, we must take a holistic view of the problem before focusing on management decisions relating to any single usage. It is also clear that economic modelling based on one or more social, economic or biological subsystems taken in isolation, will be completely misleading in assigning net values to one particular strategy or another. In this case, the interaction term which expresses the impact of any one utilization on other resources (Chapter 6) is likely to be larger than the benefits from any one potential usage taken in isolation.

Although actually calculating net benefits from different usages of the mangrove littoral is not easy to do, the most feasible approach is probably to consider for each strategy the net revenues, either per hectare of forest, or per kilometer of coastline, for any change proposed. This calculation should take into consideration the rotation time of the elements of this unique system and then integrate these over the whole area. In general, the optimal strategies of usage seem likely to be those that follow a "zoned" approach to utilization, based on a categorization of each section of the coastline in terms of its particular suitability for each potential usage. Figure 29 after Kapetsky (1985) presents a useful schematic in these cases.

The coral reef system and fisheries

The problems of fisheries management of complex ecosystems is perhaps best epitomized by coral reef resources, which Pauly (1981) suggests may contribute up to 10% of the world's annual fish catch. The extremely high diversity of these ecosystems and their very complex food webs frankly complicates the problems of their scientific management on a food web basis, although as suggested elsewhere in this document, the approach of cropping off the apical predators should probably be abandoned in favour of a strategy of fishing all trophic levels so as to keep the community 'balance' as similar as possible to an unexploited state. Paying attention to the obvious need to conserve the 'substrate' species (here referring to the wide diversity of encrusting organisms that form the bulk of the reef) is a top priority. This would mean for example avoiding destructive practises such as use of bleach or dynamite for fishing, and avoiding siltation and domestic waste runoffs in sensitive areas. This question and some suggested strategies are well described in Kenchington and Hudson (1984) who discuss a range of practical problems relating to conserving reef resources.

From the perspective of research in support of fisheries management, it is unfortunate that a great deal of the scientific literature deals with various specialized features of life cycles of coral reef fauna, and only quite recently (e.g., Munro, 1982), with fisheries management needs. This is probably inevitable, given the biological complexities of the system (see Figure 36 for a very simplified view), but does not augur well, except in qualitative way, for the use of food web information directly in management decision making, as we suggest for some simpler systems. What seems indicated is that a conservative strategy for using complex systems be adopted that equates to cautious use of all components, so that the whole ecosystem or "dissipation structure" continues to maintain its productivity. Nor is it obvious that detailed single species analysis, e.g., of the yield per recruit type, is a practical option for each of the many components of the system. Three possible approaches which take into account trophic interactions have, however, been suggested. The first, is to pick a limited number of indicator species (e.g., a herbivore, a grazer, a small predator, a detritivore, a large predator), and monitor and analyse their state of exploitation as indicators of the state of health of the ecosystem as a whole. The second, is to identify and concentrate on one or two "keystone predators" (e.g., Plectropomus leopardus on the Great Barrier Reef - Goeden, 1982). A "keystone predator" (Paine, 1969), is one which because of its economic importance, high fraction of the predator biomass, and diverse feeding habit, will integrate and reflect any adverse changes occurring at the system level. Removal of such a key predator (making up with other plectropomids of the Great Barrier Reef), about 50% of the predator biomass and 30% of the reef catch, apparently led to the wide fluctuations in biomass of other predators observed by Goeden (1982).

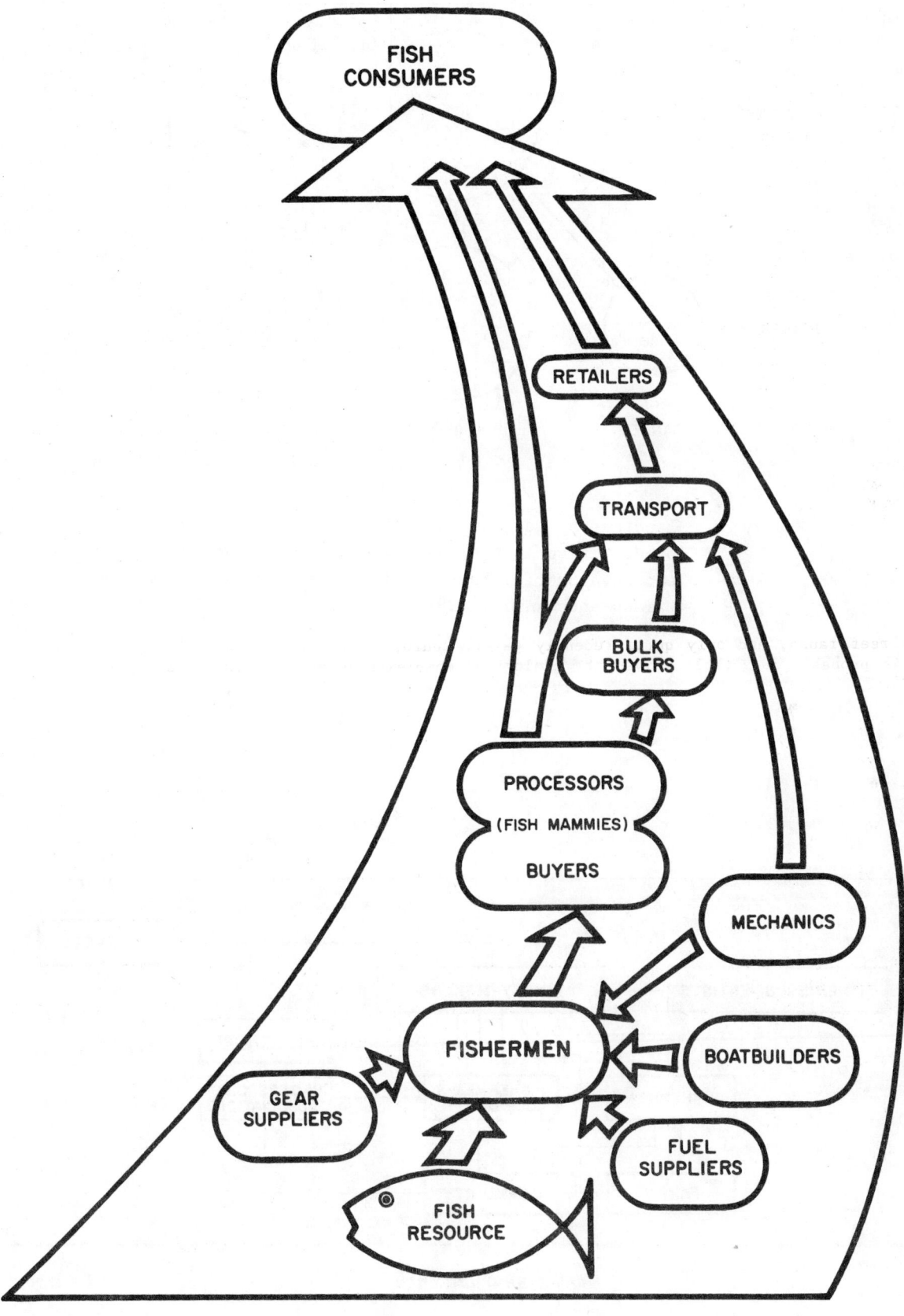

Figure 30 Structure of a traditional West African small-scale fisheries community (Redrawn from FAO/DANIDA Project Document RAF/171/DEN)

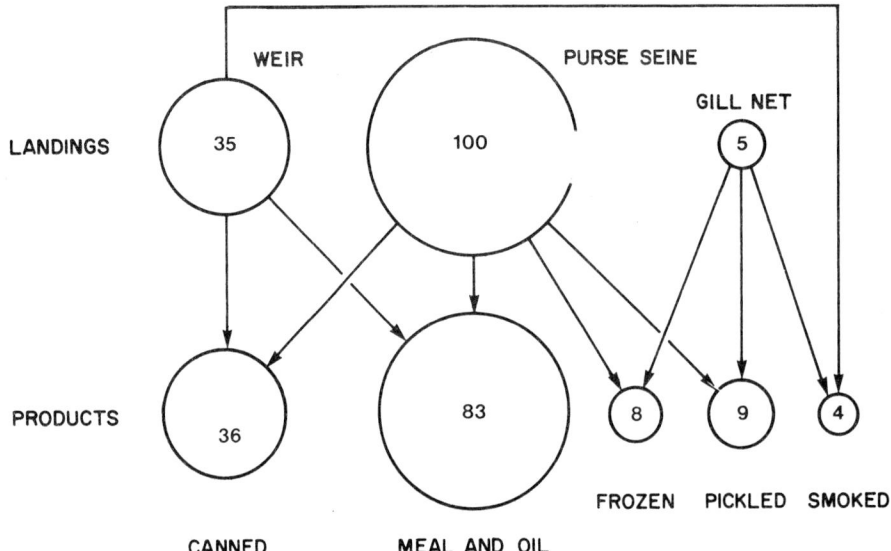

Figure 31 Landings by gear and port/markets and by eventual product type in the Bay of Fundy (New Brunswick and Nova Scotia) Herring Fisheries (Redrawn from Lamson and Hanson, 1984)

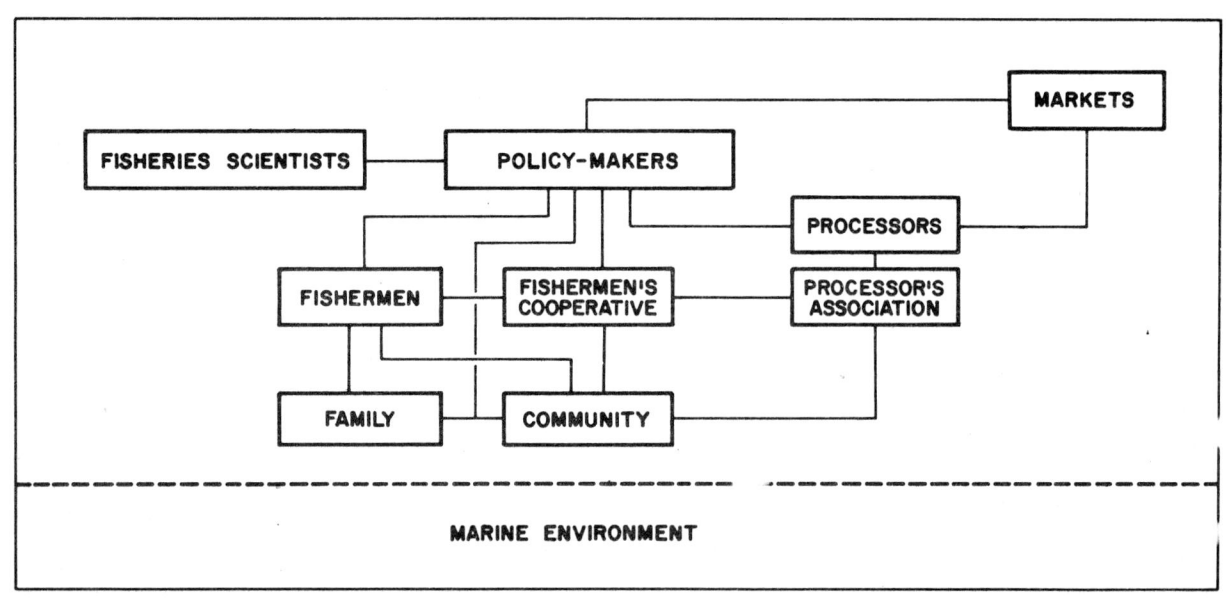

Figure 32 The structure of choice and constraint relationships in Newfoundland fisheries (Based on Lamson and Hanson, 1984)

8. DIAGRAMMATIC REPRESENTATION OF LINKAGES WITH SPECIAL REFERENCE TO FOOD WEBS

Diagrams that illustrate how components are interconnected are of great utility in understanding complex interrelationships, both in aquatic food webs, and in understanding the human components of the fishery 'dissipation structure' that results in the fish reaching consumers (see, e.g., Figure 30). Such diagrams can also be helpful in visually integrating social, economic and biological factors, such as the hierarchy of events in the secondary and tertiary components of the fisheries subsystem (e.g., Figures 31 and 32). These go beyond just those aspects of the biological production and harvesting of a fishery system.

Concentrating for the rest of this chapter on the marine ecosystem and its representation, we may note that methods of displaying species relationships in this way by means of food webs have not yet been standardized, largely because there are different levels of information content in our knowledge of species interactions, and different objectives for information transfer. Thus, as in Figure 3, we may simply find a representation of a flow of energy or material from the prey to the predator symbolized by an arrow in the direction of energy flow. It is also possible (e.g., Figure 33) to roughly quantify the biomass of each component using the sizes of the symbols (circles) surrounding them in food web diagrams, and to estimate the rate of flow of material from predator to prey by linkages of different thickness (Figure 1). Even more elaborate representations may involve a knowledge of the quantity of materials consumed, or the rate of consumption of materials, and their calorific values, organic carbon, protein or organic nitrogen content. These more elaborate representations result from quantitative (usually laboratory) experiments on the rate of feeding of predators, and a knowledge of how the energy content of prey species is subsequently partitioned (e.g., Figure 34). These latter diagrams are one of the principal tools of the study of population energetics, and have been formalized in various ways (e.g., Figure 35A) with the objective of identifying not only the direction of flow of energy and other interractions in the food web, but also its quantification in terms of an energy budget for individual species, and for the selected food web components being considered (see e.g., MacDonald, 1983).

We may mention here another type of trophic diagram often shown in the literature: the compartment model, of which Figure 33 is an example. This may be regarded as a condensed form of the food web where subcategories of the community (e.g., benthos, zooplankton, herbivores, etc.) consisting of more than one species, are grouped together for convenience. Although this is a useful simplification, it should be used with care, since the categories of species represented in for example benthos, may be highly diverse, ranging from herbivores to secondary carnivores, so that the true impact of any species may be minimized or misrepresented in the process of simplification; also, as noted elsewhere, many marine species change their effective trophic level in the course of their life history. One other requirement for their use in quantitative calculations, is that no food transfers should be concealed within any compartment. We may note here that exchange of materials between components other than directly by predation can also be represented: dead organisms (as carrion, detritus, dissolved organic matter or nutrient salts) may supply other components, often "lower down" in the web, or be "stored" for indefinite periods in biologically inactive (e.g., anoxic) parts of the environment such as in bottom sediments. Similarly, contributions from unspecified components outside the food web may be portrayed. Although the food web components modelled are usually chosen to be largely self-contained and receiving little material from outside components, as we have noted earlier, realistically this is never completely feasible. Interactions other than trophic ones may also be shown by means of a dotted line implying transfer of influence but not of material (e.g., Figure 36). Non-trophic competition between species for substrate or niches are examples of such influences.

Mention should be made here of the new field of 'loop analysis' (e.g., Saila and Parrish, 1972; May, 1974; Li and Moyle, 1981) which considers solely the functional linkages between components (species), and the direction of flow of material and influence between them in making predictions. Applications (such as in the last reference quoted), include evaluating the impact of introducing an exotic species into a food web.

The influence (positive or negative) of one organism on another can be expressed graphically by this methodology (e.g., Figure 35B); arrows indicating the direction of interactions with 'loops' connecting species in one or more closed circles. Positive and negative feedacks are identified: the latter processes tending to stable system behaviour and vice versa. The problems with the use of such an approach with other than very simple food webs seem to be threefold:

(1) without quantitative information it is difficult to distinguish important from less important linkages;

(2) problems of multiple feeding strategies cannot be handled; and

(3) determining the stability of the whole system as complexity increases is difficult.

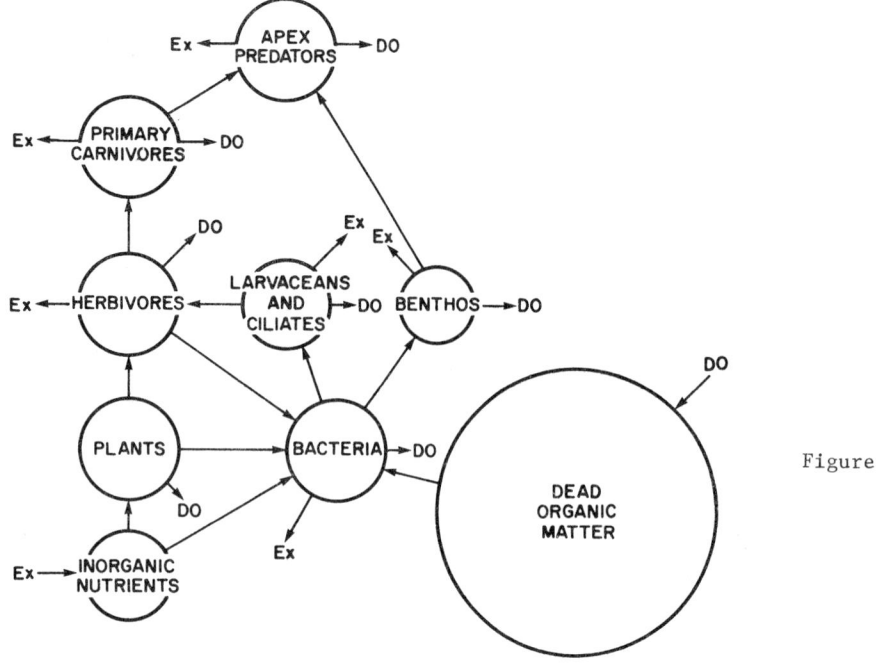

Figure 33 Ecosystem structure in terms of the nitrogen content of the main components (proportional to diameter of circles; directions of flow shown by arrows)(From Jones and Henderson, 1980)

Ex: Flow of excretory products in inorganic nitrogen pool
Do: Flow of dead material to dead organic matter pool

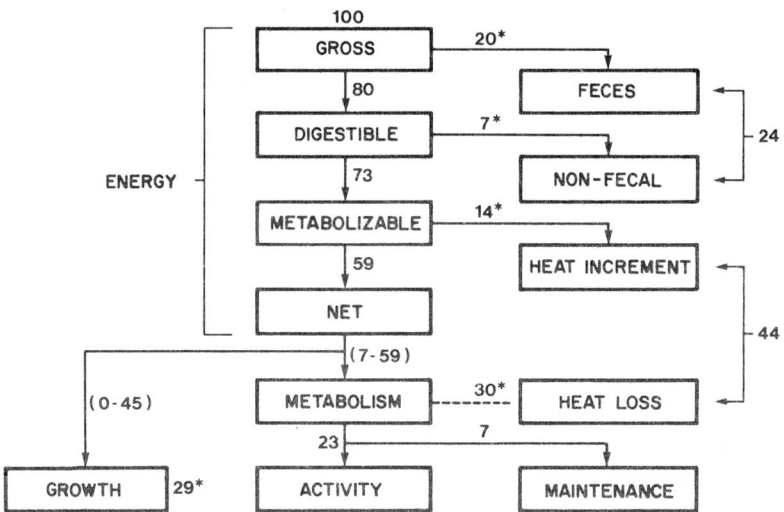

Figure 34 Average partitioning of dietary energy for a carnivorous fish. A ration of 100 calories is equivalent to about 2% dry weight/day for a 1-kg fish. Non-fecal energy is mostly excreted as ammonia and urea. $I = M + G + E$; (where I = rate of ingestion; M = metabolic rate; G = growth rate; E = excretion rate). Amounts marked with an asterisk total 100; figures in parentheses indicate the possible range of net energy distributed to metabolism and growth (Redrawn after Brett and Groves, 1979)

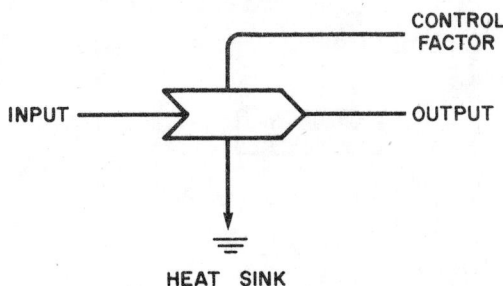

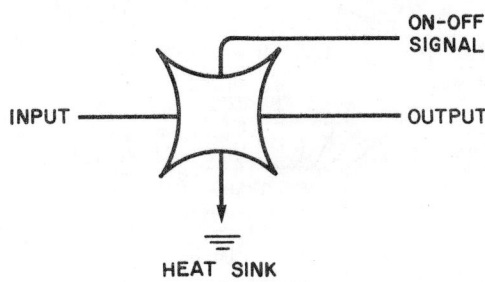

Figure 35A Some of the symbols used in the energy circuit language of H.T. Odum (Redrawn from Platt, Mann and Ulanowicz, 1981)

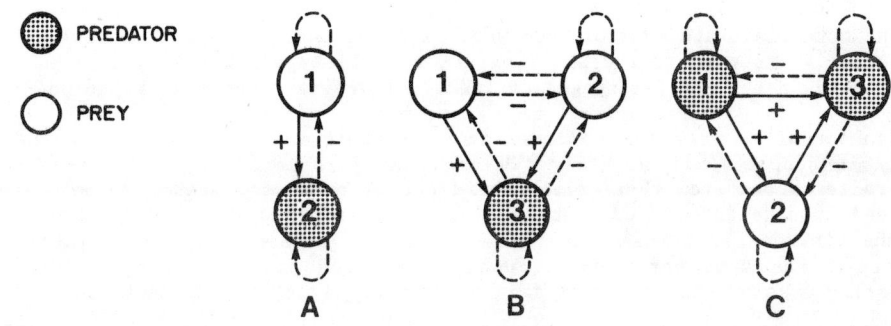

A: Stable prey-predator system

B: One predator with two preys in which one prey and the predator are self-regulating; the other prey is not

C: A system into which a destabilizing predator has been introduced

Figure 35B Loop analysis of three simple systems. Negative feedback (i.e., causing a decline in population size, or controlling the size of the component pointed to), is shown by broken lines and minus signs, while positive feedback is shown by solid lines and plus signs. Negative feedback generally results in stability, and positive feedback in increasing amplitude of oscillation (Redrawn from Li and Moyle, 1981)

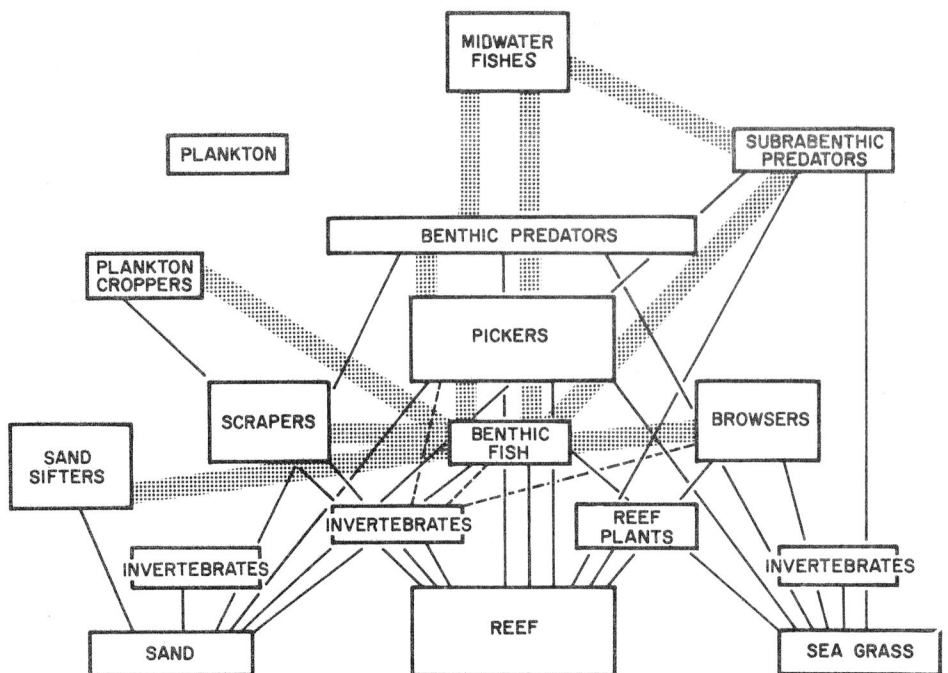

Figure 36   Summary of information flow through a community of reef fishes exclusive of food pathways. Dashed lines indicate symbiotic associations; double lines indicate space supplies; shaded lines indicate potential food competition between boxes (From Smith and Tyler, 1973)

Nonetheless, this approach points the way towards a method of at least illustrating some aspects of system behaviour in a concise visual fashion.

The more elaborate type of food web, (such as that in Figure 36 for coral reef fish) cannot be constructed without a great deal of experimental data which, for practical reasons, cannot usually be collected in the earlier stages of fishery investigations that principally concern us here.

Conceptually, it helps to integrate the ecological aspects of an exploited system with the fishery it supports, if we view both from the perspective of the common 'dissipation structure' which radiates out from those areas where energy production tends to be concentrated, and where it is passed upwards in the food web, as well as dissipated over a wider area. It is then logical to view the linkages or transactions the biomass is subjected to, from capture until it is consumed, as part of the same structure. Such an approach has the advantage that other alternative or conflicting uses of the same area can be compared directly with each other.

Interestingly enough, when the fishing industry is viewed from this perspective, the number of linkges or transactions that follow upon capture, confer added value to the product, as well as providing employment in the secondary and tertiary sections of the industry. In human terms therefore, adding to the complexity of the resource 'dissipation structure' seems to be a worthwhile activity. Figures 30-32 can therefore be viewed as components of the upper levels of the ecosystem dissipation structure and its fishery that supports this human activity.

Finally, we should note that the functional relationships between components of the fishing industry, and the constraints that apply to the industry ashore as well as at sea, also affect the fishing pressure and hence the biomass, and age and size composition of a species, as well as its resistance to competitors, ecologically speaking (e.g., Figures 40, 41 and 61). The flow of materials to man, the principal apical predator in most contemporary marine ecosystems, could thus be usefully followed through in a more or less continuous fashion from the primary production to the harvesting and consuming sectors of the fishing industry.

Portrayal of life history stages in food webs

In an ideal world, we should attempt to construct food webs that are valid for all life history stages. However, the multiple linkages (or even reversal of the direction of flow of materials) between different components in different stages of their life histories) are difficult to represent graphically. A few suggestions can be made however, which may have some value if we can regard those interactions that occur in the larval or premetamorphosis stage as separate from those occurring when a species enters its adult habitat(s), where it eventually becomes vulnerable to commercial capture. These suggestions do not belittle the importance of interactions at the larval or early juvenile stages (in fact there is growing awareness that events in the early life history control the recruitment to the adult fishable stock - see IOC Workshop Report No. 28, and the reports of the Costa Rica meeting: Sharp and Csirke, 1983, for a review of recent work on this topic). What is quite clear, however, is that:

(a) learning about species interactions in the larval, or even juvenile stages, especially in the tropics, will require quite different and very extensive commitments in research effort to elucidate them; only the first of these will be a major new problem of species identification for these early life history stages;

(b) in situations where plankton diversity is high, the biomass of larval or juvenile individuals of a given species that is included in the energy budget of a predator is likely to be almost impossible to measure, and significantly less important than the impact of the predator on the prey recruitment size;

(c) these interactions may occur over a relatively short period of time, in patchy distributions scattered over extensive areas, which makes their quantification difficult; and

(d) finally, there is a significant body of opinion suggesting that starvation, due to "mismatch" in time and space of larvae with patches of abundant food, rather than predation, may be a key factor in determining recruitment success (see however the Annex).

These points suggest that although individual (larval) prey items can contribute significantly for relatively short periods to the diet of predators, they should perhaps be regarded in many cases as being taken incidentally to generalized feeding on the planktonic community, by carnivores which extract their food from the plankton in an opportunistic fashion for most of the year.

It may be suggested as a convention therefore, that the impact of the morphologically "adult" stages (i.e., those stages occupying the adult niche(s) in the ecosystem, even if they are not sexually mature), on larval stages of other organisms, be represented by a dashed line. Thus, although adult cod feeding on herring would be represented by a solid arrow on a trophic diagram, the latter also act as predators on cod larvae between the time of cod hatching and metamorphosis, which would be represented by a dashed line. This clarifies that the concept of a food web as a way of illustrating trophic interrelationships can be usefully extended to include information flow of all kinds between participants in an assemblage and not just movement of biomass. This is also shown in Figure 36 from Smith and Tyler (1973), which is a compartment model illustrating that species may be grouped by _mode_ of feeding instead of food type; (i.e., adaptations for exploiting the same food in a different form exist, and are one of the reasons for the complexity of some marine food webs). From this figure we can see that other interactions, e.g., symbiotic associations and competition for space, may also be of importance, and can also be readily illustrated.

Compartment models attempt to simplify the complexities of food webs by including all elements occupying the same place and having the same function in the food web, in a common 'compartment', so that the same transactions are supposed to take place for elements combined in this way. Figure 37 from Jones (1982a) shows some simple examples, which in this case are intended to bring out those elements of interaction which are common to planktonic food webs. Used in this way, diagrammatic representations are obviously able to communicate a great deal of information in a very economical fashion.

The use of compartment models in simulating ecosystems has, however, led to criticisms (e.g., Mann, 1982), particularly since if individual species are condensed down into broad categories, e.g., "browsers", 'secondary carnivores' etc., this can minimize the member of linkages. This, in turn, directly affects ecological efficiency estimates and also because the time and space scales and the frequency of nutrient cycling, are so different at different levels in the system, as we have discussed elsewhere in this report.

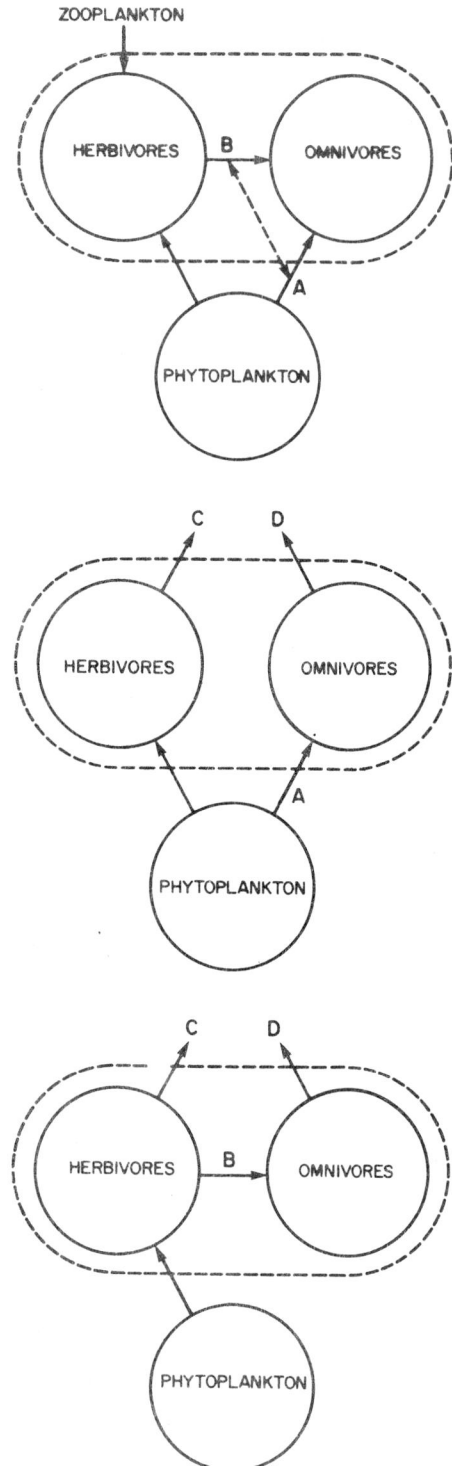

Omnivory in zooplankton
An important adaptation is omnivory. Omnivores have the capacity to switch feeding from one trophic level (A) to another (B)
Seasonal variations in omnivory could lead to significant variations in the grazing pressure on the herbivore component of the zooplankton.

Omnivores as herbivores
As long as omnivores act as herbivores (Flow A) there is a maximum of grazing pressure on the phytoplankton. This depicts a situation of maximum grazing pressure on phytoplankton and maximum tendency for zooplnkton biomass to increase.

Omnivores as carnivores
If omnivores act wholly as carnivores (Flow B) grazing pressure on the phytoplankton should be reduced to a minimum. This depicts a situation of minimum grazing pressure on phyoplankton and minimum tendency for zooplankton biomass to increase.

Figure 37 From Jones (1982a) showing how various aspects of trophic interaction can be easily represented in diagrammatic form

Despite this critique, compartment models are coming to be used for preliminary investigation of the likely influence of a change in fishing strategy on potential yields, and for organisms that do not differ too greatly in size, scale and turnover rate, they perhaps represent a reasonable way forward at present. For example, Sheridan (1984) investigated the relative impacts of partially eliminating discarded (dead) fish and small shrimp by-catch on the penaeid shrimp biomass, compared with elimination of fish and small shrimp (live) on bottom by the use of selector trawls. In attempting to weigh the loss of food for shrimp in the first case, against the effect of increased fish predation on shrimp in the second case, the model allowed some broad generalizations to be made that would certainly be of use to managers.

9. QUANTIFYING PRODUCTION WITHIN THE FOOD WEB: SOME PRELIMINARY APPROACHES

9.1 Ecological Investigation and Modelling

The field of Systems Ecology (e.g., Swartzman, 1979 and Larkin and Gazey, 1982) takes advantage of current powerful mathematical tools and computing technology to model complex ecological systems as mathematical analogues that are intended to preserve the main features of the system being modelled. After a decade or so of this kind of work, a greater understanding of the capabilities and limitations of this approach is now available, and if anything, we are led back to simpler, more robust empirical approaches to food web modelling (see Larkin and Gazey, 1982). A variety of computer models of marine food webs, or at least of trophically interlinked food web components, have been developed. Many of these are complex and are rarely usable in an 'off the shelf' fashion without considerable effort in estimating population parameters, but most of them attempt to build up a picture of the consequences of the flow of material or information at a system level, by assembling relatively simple 'building blocks' of the kinds discussed in this document. It is our contention, however, that a descriptive phase involving familiarization with the area/habitat in question and the species interactions, is an essential precondition to a sensible modelling exercise, which should then continue to develop in parallel with field observation and experiment.

9.2 Some Indices of Interaction Between Predators and Prey

It could be useful, particularly when comparing the impact of more than one predator on a common prey species, to have some idea of their relative consumption rates. If we assume for the moment that problems of availability of prey to predator are not a problem, then, an annual weighted Index of Predation by predator A on prey C could be defined as:

$$_A I_C = \sum_{season} \left[\begin{array}{c} \text{proportion prey C} \\ \text{in stomach contents} \\ \text{of predator A} \end{array} \right] \times \left[\begin{array}{c} \text{predator} \\ A \\ \text{biomass} \end{array} \right] \times \left[\begin{array}{c} \text{Food consumed as pro-} \\ \text{portion of predator} \\ \text{weight per day} \end{array} \right]$$

Obviously, the use of spatial overlap alone is inadequate, if considerations of availability such as vertical distribution (e.g., day-night migrations and predator preference/capture efficiency), considerations are not taken into account. On the other hand, a very limited degree of spatial overlap may reasonably suggest a limited degree of interaction. We may note also that although migratory predators may spend less time feeding on a prey species than 'residents', their generally higher feeding rate is important, and this is allowed for in the third term of the above expression.

An index of competition (I_c) between two predators A + B could also be postulated of the general form:

$$I_c = \left[\begin{array}{c} \text{\% similarity of} \\ \text{diet A + B} \end{array} \right] \times \left[\begin{array}{c} \text{seasonally adjusted} \\ \text{overlap of species} \\ \text{ranges for A + B} \end{array} \right]$$

Comparisons between components of the food web in this fashion should detect pairs of species most likely to be affected by changes in biomass of the other. This could be helpful in preliminary risk evaluation of the possible effects of intensive harvesting of one or more species on those associated with it trophically.

9.3 Estimating Relative Trophic Levels from Food and Physico-Chemical Studies

Food consumption patterns of marine finfish species often show major changes, both in the course of development and on a seasonal basis, in response to abundance, or more properly, to the availability and the size of various food items (Ivlev, 1961). All of these factors make construction of a food web for any given fish community a rather tenuous process, even if all food items can be identified. Of course not all components of stomach contents can be identified, due to problems particularly with fragmentary and digested organisms, and taxonomic problems and losses from the stomach on capture. Clearly, there are also problems, as we have outlined earlier in establishing the trophic level of a given organism in the system. Although the idea of the trophic level is no longer strictly necessary for the food web concept, the relative 'distance' above the base of the food web is likely to be given in a very tentative fashion judging from the size and known feeding habits of the species as determined from stomach contents. Thus, after even preliminary examination, a fish may be characterized as a herbivore (grazer), plankton feeder, benthos feeder, or small or large piscivore. This provides at least some qualitative idea of its relative position in the food web, even if such a preliminary examination does not take into account the "switching" from one prey to another in response to abundance of alternative prey, which Murdoch (1969) argued to be a mechanism for stabilizing prey population sizes.

Isaacs (1972, 1973, 1976) argued that marine food webs (disregarding birds and mammals) are largely unstructured, and are therefore not amenable to trace element concentration study. Results from studies of different coastal ecosystems and the Salton Sea (an inland salt-water lake in the Western United States that has only been in existence for 70 years, which has a very simple food web - see review in Mearns et al., 1981), tend to support this contention. However, recent very thorough quantitative studies of food habits and incorporated Cesium-Potassium (Cs/k) ratios (Mearns et al., 1981) and Carbon isotope (13C/12C) ratios (Rau et al., in press) have shown that very useful information on the mean trophic level of a component can be derived from such data.

Table 6 from Mearns et al. (1981) summarizes the results of computations of trophic levels for three species from the Southern California Bight. The % IRI value refers to the Index of Relative Importance as computed using the methods of Pinkas, Oliphant and Iverson (1971) for weighting the importance of prey species:

$$\%IRI = \%F(\%N + \%V)$$

where %F = percent frequency of occurence of prey items, %N = percent by numerical abundance of prey item; and %V = percent by volume of prey item. This is a useful weighting factor for determining each prey items relative importance in the diet of a specific predator.

Table 6

Computation of trophic levels for northern anchovy, jack mackerel, and mako shark
(based on Mearns et al., 1981)

		(1) Assumed or computed prey trophic level	(2) % IRI	(3) (1)x(2)/100	(4) Computed trophic level = (3)+1	(5) Traditional trophic level assignment
A	Northern anchovy					
	Copepod	2.0	65	1.30		
	Detritus	1.5	34	0.51		
	Phytoplankton	1.0	1	0.01		
				1.82	2.82	II ⟶ III
B	Jack mackerel					
	Copepods	2.0	52	1.04		
	Unidentified crustacean	2.0	24	0.48		
	Unidentified matter	1.5	8	0.12		
	Unidentified fish	3.0	8	0.24		
	Polychaetes	2.5	5	0.13		
	Squid	3.06	1	0.03		
				2.04	3.04	III
C	Mako shark					
	Pacific mackerel	3.54	65	2.30		
	Jack mackerel	3.04 [a/]	19	0.58		
	Unidentified fish	3.0	17	0.51		
				3.39	4.40	IV ⟶ V

[a/] From above

Table 7 from Mearns et al. (1981) shows results from the combined feeding and Cs/K ratio studies for 22 species of pelagic animals from either the Southern California Bight (SCB) or the Eastern Tropical Pacific (ETP). The log transformation of the Cs/K ratios correlates significantly with the computed or assumed values of trophic levels in Table 7, leading the authors to report an average factor of increase of about 2.3 to 2.4 times in the Cs/K ratio at each trophic step. There is a continuum of trophic levels from the assumed levels for zooplankton (of 2) up through the computed value of 5.02 for <u>Carcharadon carcharius</u>, the white shark.

Table 7

Summary of locations of capture, Cs/K ratios, assigned trophic levels (assumed or computed) and conventional expression of assigned trophic levels for 22 species of pelagic animals from the southern California Bight and the eastern tropical Pacific, 1978 and 1980
(based on Mearns et al., 1981)

Species No.	Common name	Location[b]	$Cs/K \times 10^{-6}$	Assigned trophic level	Conventional trophic level
HERBIVORES (II)					
1	Coastal zooplankton	SCB	< 2.07	2.00[a]	II
2	Oceanic zooplankton	ETP	3.30	2.00[a]	II
PRIMARY CARNIVORES (III)					
3	Northern anchovy	SCB	< 1.86	2.82	II ⟶ III
4	Blue whale	SCB	11.00	3.00	III
5	Flying fish	ETP	7.00	3.00	III
6	Pacific sardine	SCB	4.02	3.01	III
7	Market squid	SCB	2.39	3.05	III
8	Jack mackerel	SCB	5.73	3.04	III
INTERMEDIATE (PRIMARY-SECONDARY) CARNIVORES (III-IV)					
9	Squid	ETP	1.94	3.52	III-IV
10	Pacific mackerel	SCB	7.23	3.54	III-IV
11	Frigate tuna	ETP	8.90	3.56	III-IV
SECONDARY CARNIVORES (IV)					
12	California barracuda	SCB	4.20	3.74	III ⟶ IV
13	Pacific bonito	SCB	8.01	3.80	III ⟶ IV
14	Thresher shark	SCB	24.0	3.82	III ⟶
15	Swordfish	SCB	12.3	3.97	IV
16	Blue shark	SCB	13.2	4.00	IV
17	California sea lion	SCB	10.7	4.02	IV
INTERMEDIATE (SECONDARY-TERTIARY) CARNIVORES (IV-V)					
18	Yellowfin tuna	ETP	12.7	4.23	IV ⟵ V
19	Skipjack tuna	ETP	8.59	4.30	IV-V
20	Mako shark	SCB	19.7	4.39	IV-V
21	Silky shark	ETP	22.8	4.55	IV-V
TERTIARY CARNIVORE (V)					
22	White shark	SCB	31.7	5.02	V

b/ SCB = Southern California Bight and adjacent waters; ETP = Eastern Tropical Pacific (oceanic)

c/ Assumed trophic level; all others computed from stomach contents data according to method described in text

Mearns et al., (1981) also report that these ratios do not increase at the same rate for inshore and offshore, or benthic food webs. They also recommend that the dispersion of trophic levels of prey items be recorded, which would indicate the observed trophic spectra each species feeds upon. Such studies, which seem to offer promise for the future, support the concept that variations in time and space are to be expected in the way food webs are structured. Evidently if the methodology described by these workers proves easy to apply, it may provide a further lease of life for the trophic level concept, and for the first time, an objective way of determining it directly. It seems unlikely in the short term, however, to provide a cheap and effective field procedure.

Employing similar logic in studying the 13C/12C ratios in relation to the samples obtained in the Mearns et al. (1981) studies, Rau et al. (in press) report that the 13C/12C ratios also increase with trophic level between the Coastal California and equatorial Pacific waters, and recent work by Mills, Pittman and Tan, 1982 tend to confirm that the "distance" of a given species above the level of primary (plant) production may well be measurable in this fashion. In a parallel sense, it is important to note that some organic pollutants and naturally occurring heavy metals (e.g., mercury in swordfish) also accumulate progressively with higher trophic level and increasing mean age of fish.

9.4 Production, Fishing and Natural Deaths

The production by each species in a food web over a period of time can be defined in a number of ways, which have been summarized by Allen (1971). In simplest terms, if P_n is the net production of species n; F_n the fishing mortality rate it is subject to; and M_n its natural mortality rate, then, using a simple numerical approach, following Dickie (1972), we can:

1. Write $P_n = \bar{B}_n Z_n$ where Z_n is the overall species-specific mortality rate, comprising $F_n + M_n$, and $\bar{B}_n$ is the mean biomass of the organism in question. The fraction of P_n harvested by man (the yield) is defined by:

$$Y_n = (F_n/Z_n) P_n = \bar{B}_n F_n$$

The fraction consumed by other members of the ecosystems is similarly defined by:

$$D_n = (M_n/Z_n) P_n = \bar{B}_n M_n$$

2. Alternatively, we could define production from the trophic efficiency of a predator population and write:

$$P_n = K_n I_n$$

where K_n is the overall food conversion efficiency of the predator, and I_n is the rate of food intake of the predator population. Of course each prey item has a specific conversion coefficient which will need to be estimated with this approach, entailing a considerable amount of experimental work. Nonetheless, a great deal of this work has already been done, so that the variable a_n (the food consumption rate per unit biomass of predator n), has been documented for a wide variety of species (see e.g., Conover, 1978), and we note here, what will be amplified later, namely that I_n is defined by:

$$I_n = a_n B_n$$

3. We could also define production of a predator n+1 in terms of the production of its prey(s) in the trophic level (n) below. Thus, for obligate predator-prey populations in which all prey mortality is due to predation, if food conversion efficiencies of predator + prey can be considered roughly the same:

$$\frac{P_{n+1}}{P_n} = \frac{K_{n+1} I_{n+1}}{K_n I_n} = \frac{I_{n+1}}{I_n}$$

this leads to:

$$\frac{P_{n+1}}{P_n} = \frac{a_{n+1} B_{n+1}}{a_n B_n} < \frac{B_{n+1}}{B_n}$$

Since, as we note elsewhere, $a_n > a_{n+1}$, food consumption per unit body weight will tend to decline with increased size of organism. The ratio of production at two adjacent trophic levels should be less than the ratio of biomasses, because not all of the biomass of prey will be consumed by the predator at the next trophic level. However, for other than "stratified pelagic systems", the proportion of prey species biomass that is consumed by predators is probably rather high for adult fish in marine food webs, even though the particular predator may not always be of commercial importance or even appear eventually in commercial catches. There are obvious dangers with this approach of comparing directly the biomasses of two vertically linked components of an ecosystem to determine predation rate, if not all of the trophic linkages are known, and some more recent concepts have called into question some of the earlier predictions of "simple" trophic concepts as outlined, for example, by Lindemann (1942).

Classically, Lindemann (1942) envisaged animals as belonging to definite trophic levels, thus giving rise to a pyramid (Figure 2): the biomass falling off with increasing trophic level. Much recent research has shown that this need not be the case, given (Isaacs, 1972, 1973) that we have three types of transfer of energy or matter:

k_1: conversion of food to living tissue

k_2: "loss" of food energy by respiration

k_3: conversion of food into a non-living but useable form - e.g., detritus

where $k_1 + k_2 + k_3 = 1$

Under these circumstances, it is possible for predator biomass to be similar to or, even greater than, the biomass of herbivores. For example, if $k_1 > k_2$ for predators, and/or $k_3 > k_1 + k_2$ for herbivores; i.e., if conversion of food into living tissue by predator is efficient (say around 30%), respiration rate relatively low and/or the production of detritus, especially by herbivores, is high so that a high biomass of detritivores exists (which are also fed on by predators), (Platt, Mann and Ulanowicz, 1981), it is possible that the biomass of predators can be high relative to the biomass of herbivores, which is unlike the situation predicted by the simple model. There seems evidence for this situation in some pelagic and benthic ecosystems.

Incidentally, the Isaacs model assuming an unstructured food web, yields reasonable predictions in some other circumstances, and lends some support to the use of compartment models, at least for representing the energetics of ecosystems.

Extension of simple production modelling to multispecies systems: some pitfalls and problems

As noted by May et al. (1979), "A prey-predator system cannot be managed by applying MSY notions to each species individually". More generally, from the previous sections, it should be clear that the yield to be expected from a given single species fishery, whether under "equilibrium" conditions or otherwise, cannot be entirely dissociated from the impact on the same stock of the abundance of its predators or prey. The big difference is that usually the natural mortality rate of, and predation on, smaller often immature individuals is usually highest, while the selective properties of most fishing gear is aimed at the larger, often mature, "commercial" sizes (Figure 38).

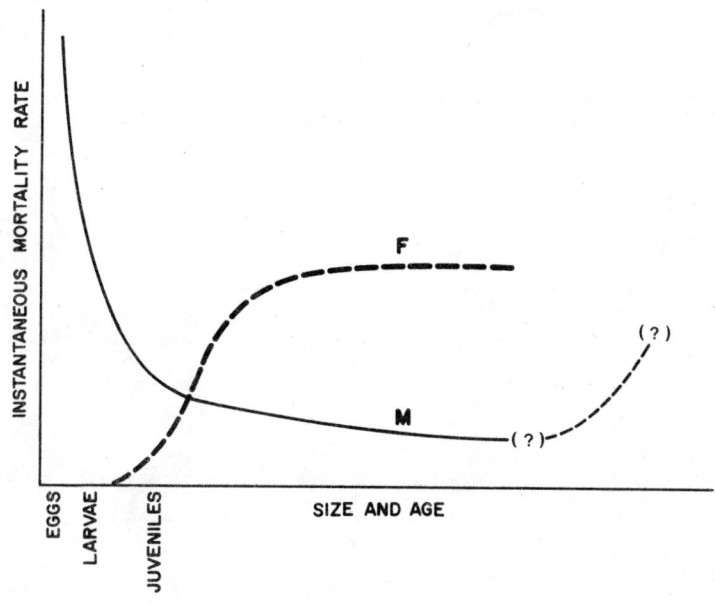

Figure 38 Showing how fishing mortality and natural mortality rates for most commercial fish species caught in area-swept gear (e.g., trawls) have quite different trajectories with size and age. (NB: Survival rates for old individuals may decrease, thus showing departures from this generally assymptotic situation)

The consequences of this are not totally evident, except that the recruitment overfishing impact on large, mature members of a stock will likely be more serious than predation on the possibilities of stock replenishment, for the same number of individuals removed. In the simple case however, we can maintain the additive nature of exponential mortality rates postulated in fisheries population dynamics theory (e.g., Ricker, 1975), where total mortality rate $Z = M + F$.

Some preliminary approaches to modelling catch and effort data from several species in a fishery in combination using the logistic model, have been made, e.g., by Pope (1980), but without taking into account trophic interaction. Such interactions must be important however, and one simple way of visualizing them is provided by extending production modelling theory to include a simplistic formulation of the total production from a species, and its partitioning into fishing yield and natural deaths due to predation. Two interesting features of production curves for a stock under these assumptions are shown by Figure 39 from Caddy and Csirke (1983), namely that as effort and mortality increases towards conditions when MSY (Maximum Sustainable Yield) is being extracted, the total biological production is likely to have also increased, and then fallen with further increases in effort as MSY conditions were approached, unless of course, the total production curve and the yield curve are quite different in shape. If M is high for a stock, this decline in production is likely to begin early in the fishery, to the point that for short-lived species with a high M, yield may not be sustainable, or only at fairly low ratios (F/Z) of exploitation, if predator populations remain high. (Biologically, we may consider such stocks as already significantly stressed by predation to the point of losing some of their "resilience".) Also the logistic and other production models are of rather uncertain relevance once MSY conditions have been significantly exceeded, and yield becomes more heavily dependent on recruitment and on the ability of a species to maintain to competitive position in the ecosystem at low stock sizes. We should also note (see later) that exploitation of the predators on this stock will influence potential yield of its prey (Figures 40 and 41), although rarely to the extent predicted by simple food webs which assume static linkages.

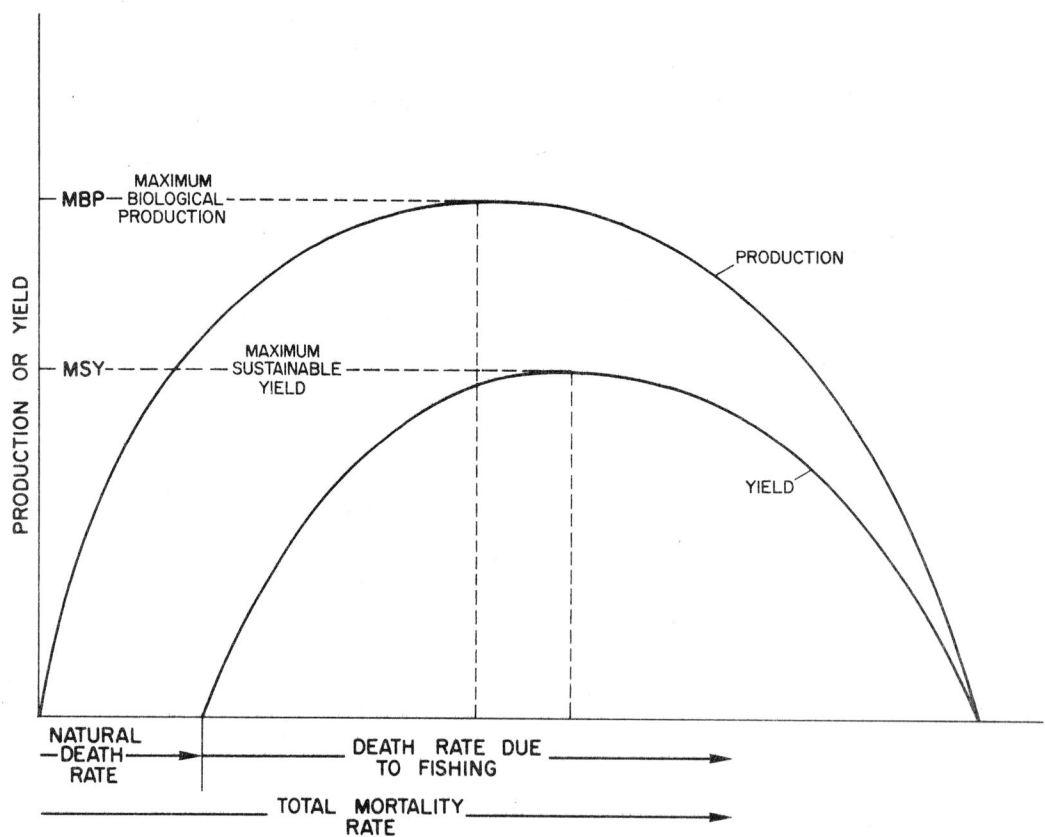

Figure 39   A simple hypothesis as to the response to fishing of total production and yield under the logistic assumption (From Caddy and Csirke, 1983)

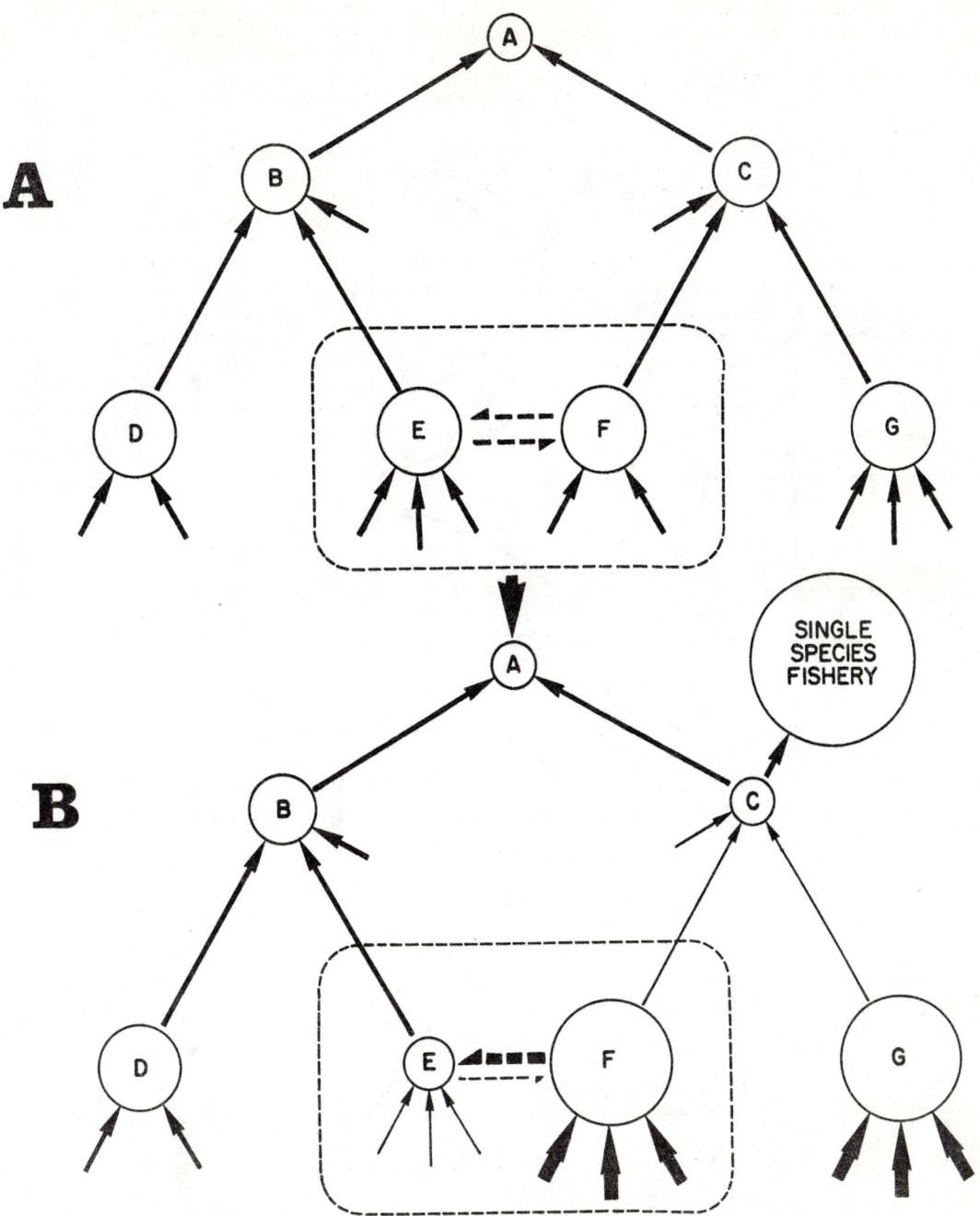

Figure 40   Predator-prey interactions and exploitation (from Caddy, 1984)

(A) Simple unexploited food web for species A-G; solid arrows show flow of biomass from predator to prey with diameter of circles corresponding to species biomasses. (Solid arrows without origin correspond to unspecified prey species, and dashed arrows inside dotted outlines indicate competition between species E and F for space, food, etc.)

(B) A monospecific fishery for species C is now in operation. The thickness of arrows is proportional to rate of flow of material from prey to predator, or (if the arrows are reversed), the degree of control exerted by predator on prey.

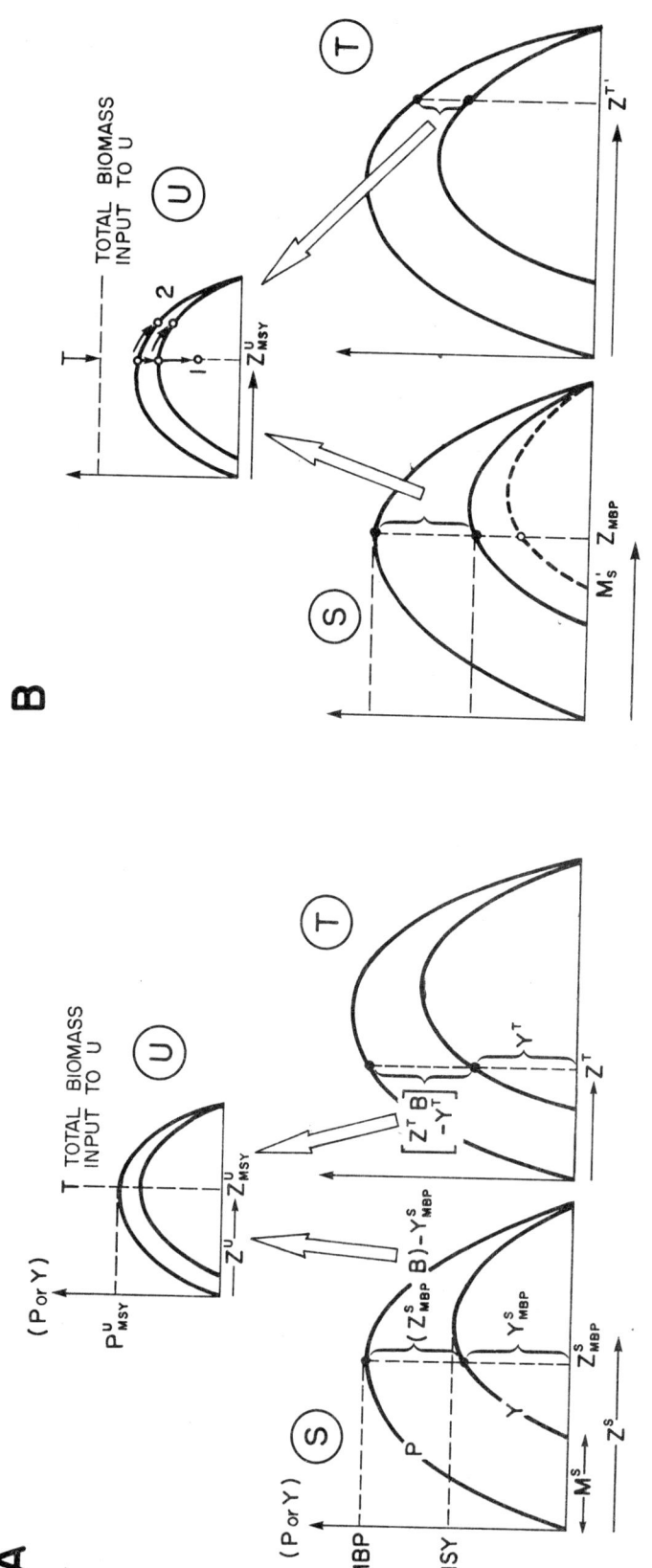

Figure 41 Showing how simple models of total production and fishery yield may be linked to illustrate (in this case), the impact of an increase in effort on one species (T) of a trophically-linked group of commercially exploited species

Another point of this simple model that seems to correspond to reality is that the production from the whole system (yield plus predation) is likely to reach a maximum to the left of MSY - in effect somewhere close to the point of Maximum Economic Yield (MEY) as defined by fisheries economists (from different criteria), or near to the point of $F_{0.1}$ (now used in many stock assessments to give a fishing intensity less than f_{MSY} or F_{MAX} Gulland and Boerema, 1973). Preoccupation with fishing prey (forage fish) populations at or around MSY has been expressed, e.g., by May et al. (1979), who note that "preservation of the ecosystem would seem to require that stocks not be depleted to such a level that the populations productivity or that of other populations dependent on it, be significantly reduced". A somewhat similar concept is implied at the single species level by F_{MBP} (Caddy and Csirke, 1983), which is the fishing mortality rate under logistic assumptions that Maximizes Biological Production (MBP) from a given species. These ecologically-based approaches to resource management are now embraced by at least one international convention for management of marine resources, (that of the Commission for the Conservation of Antarctic Marine Living Resources (CCAMLR) (Table 8).

Table 8

Extracts from the Convention on the Conservation of Antarctic Marine Living Resources

---

ARTICLE I

1. This Convention applies to the Antarctic marine living resources of the area south of 60° South latitude and to the Antarctic marine living resource of the area between that latitude and the Antarctic Convergence which form part of the Antarctic marine ecosystem.

2. Antarctic marine living resources means the populations of finfish, molluscs, crustaceans and all other species of living organisms, including birds, found south of the Antarctic Convergence.

   The Antarctic marine ecosystem means the complex of relationships of Antarctic marine living resources with each other and with their physical environment.

ARTICLE II

The objective of this Convention is the conservation of Antarctic marine living resources.

For the purposes of this Convention, the term "conservation" includes rational use.

Any harvesting and associated activities in the area to which this Convention applies shall be conducted in accordance with the provisions of this Convention and with the following principles of conservation:

(a) prevention of decrease in the size of any harvested population to levels below those which ensure its stable recruitment. For this purpose its size should not be allowed to fall below a level close to that which ensures the greatest net annual increment;

(b) maintenance of the ecological relationships between harvested, dependent and related populations of Antarctic marine living resources and the restoration of depleted populations to the levels defined in sub-paragraph (a) above; and

(c) prevention of changes or minimization of the risk of changes in the marine ecosystem which are not potentially reversible over two or three decades, taking into account the state of available knowledge of the direct and indirect impact of harvesting, the effect of the introduction of alien species, the effects of associated activities on the marine ecosystem and the effects of environmental changes, with the aim of making possible the sustained conservation of Antarctic marine living resources.

The simple approach to biological production modelling implied in Figure 39 in theory could be extended to several species, for example, to a several species Assemblage Production Unit or APU (Figure 41). Here, components of a simple two- prey one- predator system are each considered to be modelled by the logistic model as above. This example illustrates, however, that some features of such a "trophically linked" system may have properties not shown by the individual species. For example, consider Fig. 41, where prey species S is being fished at the point of Maximum Biological Production (MBP), prey T at a lower intensity than this, and predator U at F_{MSY}:

Only a proportion of the biological production of the two prey species S and T, namely $P_u = \sum_{i=1}^{2} (Z_i B_i - Y_i)$, is supporting the production of predator U.

The apparent paradox occurs when you consider what happens when the fishing intensity on species T is increased, thus reducing the production from species T which is available to predator U, while all other parameters ostensibly remain the same. It soon becomes evident (Figure 41B) that it is not possible to maintain all parameters constant under these circumstances. Changing the point on the production curve for species T must also necessarily change the production curve itself for at least one of the other components, and probably both.

The managers of the fishery on U (and species U itself) appear then to have three options:

1. Production and yield for species U (and S) may only be maintained constant if U makes up for lost production from species T by switching its feeding to one or more new food items not shown in Figure 41. This is a frequent strategy for many species, especially seasonally, but may not always be entirely successful in maintaining the same level of predator production.

2. If we specify that total predator mortality ($_u Z_{MSY}$) remains the same, and the food requirements are also to remain constant, consumption of alternate prey species S by predator U may have to increase if production of prey T is reduced by fishing. Presumably this will also increase the natural mortality rate of S from M^S in Figure 41A to M'_S in Figure 41B. This drops the potential yield of S, since the overall yield to fishing (and, in effect, the exploitation rate on S), is reduced.

3. Presumably, if strategies 1 and 2 above are not possible, the fishery on U will either produce a lower yield for the same effort (point numbered 1 in Figure 41B),

or:

4. Effort will increase in response to reduced abundance of U until point number 2 in Figure 41B is reached, where population size and hence food demand of U are reduced.

No final resolution between these possiblities can be generalized upon in such an oversimplified scenario, except to note that a series of continual adjustments of this type must be going on in practise; contributing to the "noise" or even structural instability of many fisheries, especially for species lower in the food chain, and the common departures from "equilibrium production levels" that are frequently noted.

In discussing the quantitative approach to food web analysis, Ulanowicz (1980) notes that four types of flows need to be measured (our qualifications in brackets):

(1) exchanges between (recognized) compartments within the system;
(2) inputs from sources (defined as) outside the system;
(3) usable exports to outside the system (as defined);
(4) dissipation of materials into a form of no use to any system.

The consequence of these points for productivity analysis is that it is probably not valid in most cases to estimate productivity of the food chain in the fashion seen in earlier texts, where "bottom-up" calculations were based on an overall estimate of primary production, to which were applied successively loss rates for herbivores, primary carnivores, secondary carnivores etc., of roughly 90% to eventually evaluate potential fish yield. Clearly, this type of calculation is incorrect in several ways:

(1) the estimates of biomass lower down in the food chain (especially for primary predators) are likely to have wide margins of error;

(2) the efficiency of transfer of biomass is not a constant;

(3) marine species normally move upwards in trophic chains in ontogeny;

(4) leakage occurs to undefined species;

(5) the effects of detrital recycling is not taken into account (see e.g., Newell, 1984).

There doesn't seem to be any escape from the conclusion that the main components of the system need to be identified individually, and the actual food linkages, and where possible, the order of magnitude of biomasses and transfers, need to be estimated in order to use a subset of the food web as a basis for intelligent hypothesis and experimentation.

Simple trophic analysis of a food web

As a first approximation, it may be convenient to proceed directly from a preliminary prey-predator contingency table (e.g., Fig. 39A) to construction of a tentative food web (Figure 39B), even though estimation of the rate of feeding of predators in absolute terms requires considerable knowledge of ingestion, digestion or egestion rates, both as a function of environmental factors, and of type and abundance of food. These latter factors will be touched on later, but for the moment the assumption is made that the relative volume of identifiable food contents per predator body weight, is an index of their fraction in the total diet. This assumes that the sample of stomachs is properly weighted both for diurnal and seasonal factors, and for the distribution and relative abundance of the different size groups under investigation.

To some extent a preliminary trophic classification of this kind is possible by reference to the literature, although information is surprisingly sparse on the diets of many marine fish species.

"Stomach content analysis" (including intestinal contents) is the principal method available to the field worker with limited facilities for intensive laboratory or field studies. This methodology and the data it generates has been inadequately utilized to date, largely we believe because of uncertainties in the interpretation of the large volume of data that can be rapidly obtained by this method. To a large extent this is also because methods of population analysis until recently have placed inadequate emphasis on multispecies approaches and species interactions.

One suggested approach to analyzing the preliminary food web is given below in five steps, illustrated by means of a simple hypothetical example. It is assumed here, as a simplifying assumption, that the food requirements of a species are homogenous. In reality, it would be necessary for most species as we have seen, to divide each up into two or more size or age categories corresponding to the different trophic levels occupied in the life history; (or treat each of these as separate groups of organisms):

Step 1: Preparation of primary contingency tables of stomach contents for the main predators showing mean weight (volume) of prey in the stomach as a fraction of predator body weight (Figure 42).

		PREDATOR			
		1	2	3	4
PREY	1	0.024	0.030	0.030	0.040
	2			0.050	0.020
	3				0.019
	4			0.001	0.080
UNIDENTIFIED FOOD		0.003	0.030	0.040	0.031

Figure 42 Matrix of food consumption data for four trophically linked species showing stomach contents as percentages of body weight of the four components

Step 2: A preliminary food web is constructed (Figure 43).

Step 3: The proportion of identifiable stomach contents per item is calculated.

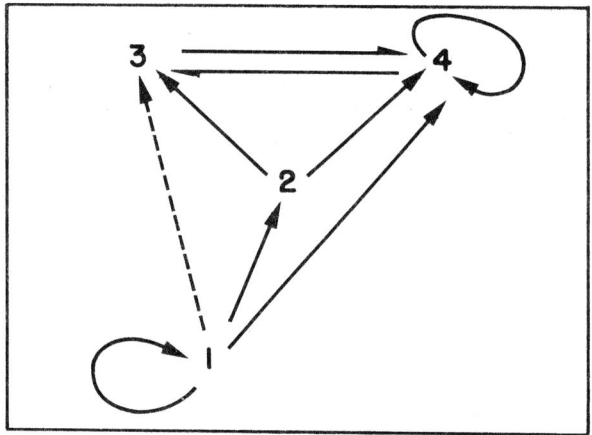

Figure 43   Simplified food web based on Figure 42

	Predator 1	2	3	4
1	0.889	0.500	0.250	0.211
2			0.417	0.105
PREY 3				0.100
4			0.008	0.421
	0.111	0.500	0.325	0.163

In matrix form

$$\begin{bmatrix} .889 & .500 & .250 & .211 \\ 0 & 0 & .417 & .105 \\ 0 & 0 & 0 & .100 \\ 0 & 0 & .008 & .421 \end{bmatrix}$$

(Unidentified stomach contents either are not included in the matrix, or allocated proportionally to recognizable items).

Step 4: Correction for feeding, egestion or digestion rate, (if possible by post-multiplication of the above matrix by a diagonal matrix B containing the estimated feeding rates):

$$
\overset{A}{\begin{bmatrix} .889 & .500 & .150 & .211 \\ 0 & 0 & .417 & .105 \\ 0 & 0 & 0 & .100 \\ 0 & 0 & .008 & .421 \end{bmatrix}} \times \overset{B}{\begin{bmatrix} 25 & 0 & 0 & 0 \\ 0 & 16 & 0 & 0 \\ 0 & 0 & 12 & 0 \\ 0 & 0 & 0 & 4 \end{bmatrix}} =
$$

$$
\begin{bmatrix} .889{\times}25 & .500{\times}16 & .150{\times}12 & .211{\times}4 \\ 0 & 0 & .417{\times}12 & .105{\times}4 \\ 0 & 0 & 0 & .100{\times}4 \\ 0 & 0 & .008{\times}12 & .421{\times}4 \end{bmatrix} = \overset{C}{\begin{bmatrix} 22.23 & 8 & 1.8 & .84 \\ 0 & 0 & 5 & .42 \\ 0 & 0 & 0 & .40 \\ 0 & 0 & 0.10 & 1.68 \end{bmatrix}}
$$

The last matrix is then our best estimate of the food consumption per unit time per unit of predator body weight.

<u>Step 5</u>: Absolute food consumption can then be obtained per unit time by post multiplication of the above matrix by a diagonal matrix containing the biomass of each predator:

$$
\begin{bmatrix} 23.23 & 8 & 1.8 & .84 \\ 0 & 0 & 5 & .42 \\ 0 & 0 & 0 & .40 \\ 0 & 0 & 0.10 & 1.68 \end{bmatrix} \times \begin{bmatrix} 100\,000 & & & \\ 0 & 15\,000 & & \\ 0 & 0 & 3\,000 & \\ 0 & 0 & 0 & 4\,000 \end{bmatrix} =
$$

$$
\begin{bmatrix} 2\,223\,000 & 120\,000 & 5\,400 & 3\,360 \\ 0 & 0 & 15\,000 & 1\,680 \\ 0 & 0 & 0 & 1\,600 \\ 0 & 0 & 300 & 6\,720 \end{bmatrix}
$$

This can be condensed further to estimate absolute food consumption by each predator per unit time T as a horizontal vector:

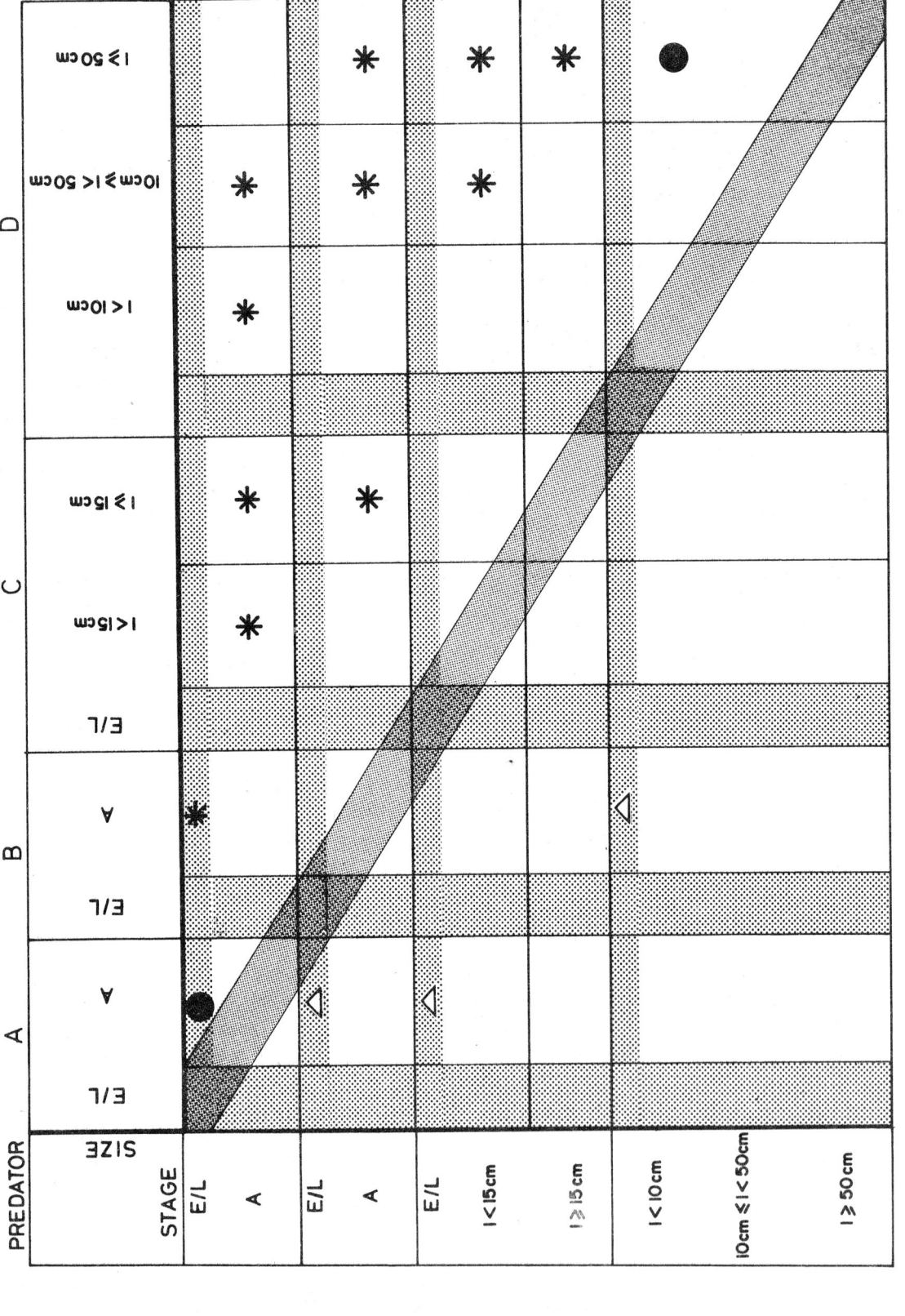

Figure 44 Prey-predator contingency table for part of a hypothetical neritic food web, including two small pelagic species (A, B) for which eggs and larvae (E/L) are recorded separately from adults (A), and two higher trophic-level predators (C, D) for which eggs and larvae and two or three trophically relevant size categories are recorded separately. Predation is shown by an asterisk; cannibalism by a solid circle. Note that reciprocal consumption of the larvae of adult predators by small pelagic "prey" falls below the matrix diagonal (triangles)

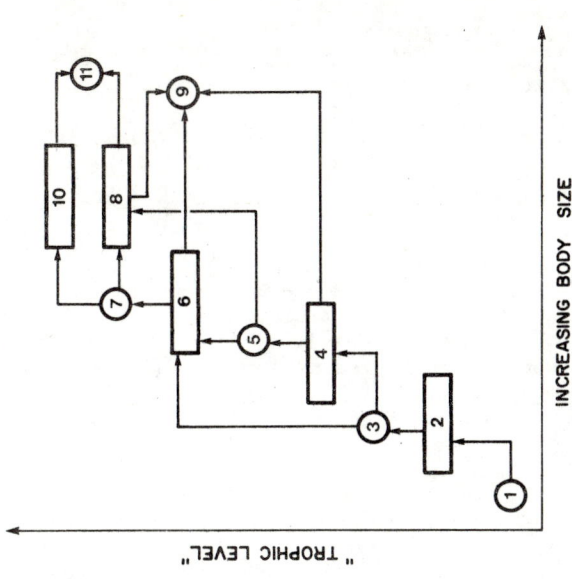

Figure 46 A food chain showing the strategy of calculation suggested by Borgmann (1983), which postulates "hypothetical species" (circles) that are effectively of constant size and are used to simplify the mathematics of computation of production when real species (rectangles) actually vary in size

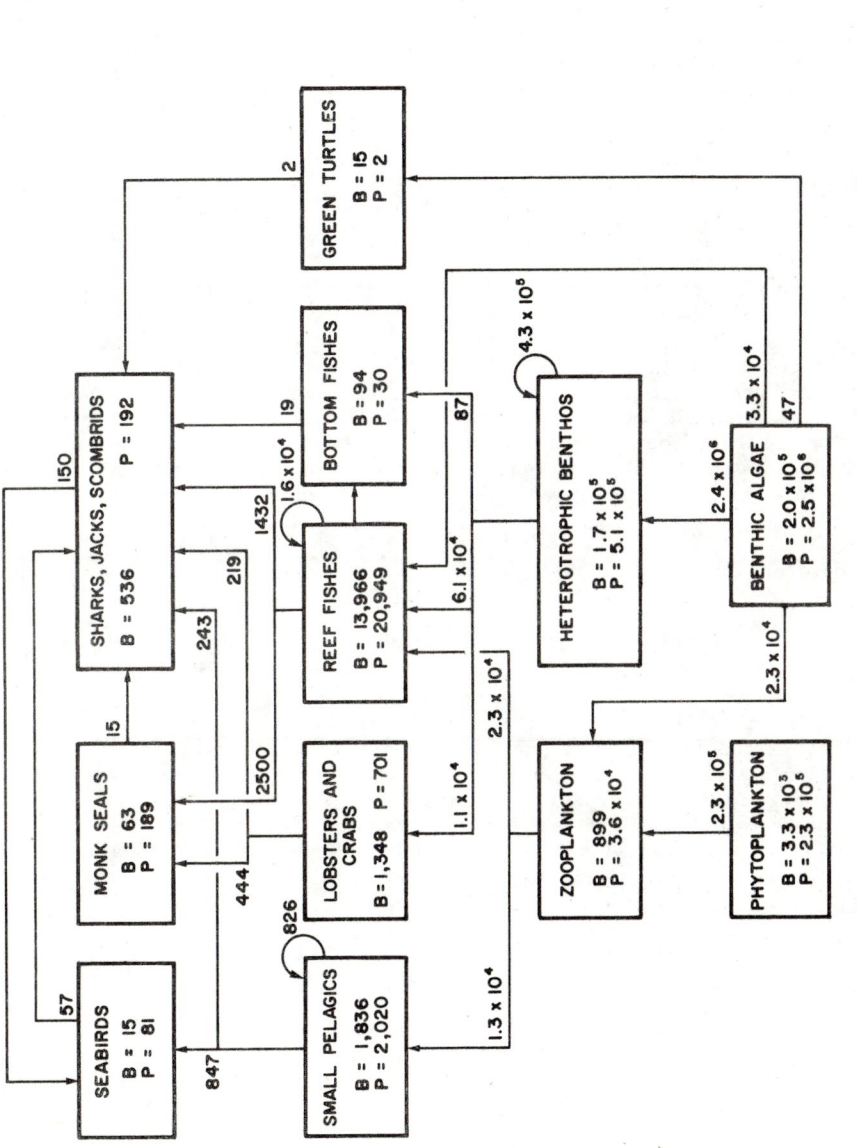

P = annual production; B = mean annual biomass (kg/km^2); values with arrows are amounts consumed of lower trophic level production

Figure 45 Flow of materials through a marine ecosystem (1 200 km^2 in area) in the north-west Hawaiian Islands (Redrawn from Polovina and Ow, 1985)

Predator

	1	2	3	4
	2 223 000	120 000	20 700	13 360

or total amount of each prey consumed by a vertical vector:

$$\begin{bmatrix} 2\ 351\ 760 \\ 16\ 680 \\ 1\ 600 \\ 7\ 020 \end{bmatrix}$$

One aspect of marine food webs that is not readily represented by the above simple matrix approach needs mention here, and is illustrated in Figure 44, namely the fact that there are often significant changes in trophic preference in the course of the life history. A modified contingency table such as that shown may then be useful for representation of different life history stages in the context of trophic interactions.

Simple Ecosystem Models

The ECOPATH model is one of a number of simple compartment or box models which attempt to simplify the complexities of massive food webs by placing food web components into major categories with known linkages (see Figure 45 from Polovina and Ow, 1985), for each of which estimates of total production, predator biomass and catch are available. The model requires that equilibrium conditions exist, when estimates of production (such as those measures suggested by Allen, 1971), and constant species-specific values of M and F derived from population assessments also apply. The importance of the assumption of equilibrium is that the biomass budget equation (for all prey species i = 1, 2, 3...n) can then be simply defined as:

$$\begin{bmatrix} \text{Biomass} \\ \text{production} \\ \text{of species i} \end{bmatrix} - \begin{bmatrix} \text{Predation} \\ \text{on} \\ \text{species i} \end{bmatrix} - \begin{bmatrix} \text{Non-predatory} \\ \text{mortality} \\ \text{on i} \end{bmatrix} - \text{Catch}_i = 0$$

and then used to estimate unknowns. Estimates of the rate of flow through the food web are only strictly valid under these circumstances, (as of course, is the assumption of constant natural mortality rates). This poses some serious problems for the general applicability of such modelling approaches, but this type of model still seems to offer a first entry into what is, as we have seen, a complex multispecies 'landscape'. One other problem faced by this approach is the difficulty of dealing with intraspecific variations in diet with season and size. One approach that offers a partial solution to this problem is suggested by Borgmann (1983) where size ranges of species are extended, namely, to postulate hypothetical species of a fixed size that allow the some general increase in body size with distance upward in the food web to be treated more easily (Figure 46).

It is not apparent, however, that this device allows for those changes in size range that occur upon exploitation.

10. SIMPLE CATEGORIZATION OF LIFE HISTORY STRATEGIES: r- AND K- SELECTION AND THE ECOLOGICAL NICHE

The useful, if rather over-simplified concept of the r-K continuum of life history strategies, first coined by MacArthur and Wilson (1967), has proved valuable in attempting to integrate life history information in relation to the role of a species in the marine ecosystem, even if it is difficult to quantify in practical terms. Stated simply, it is postulated that two major evolutionary strategies can be recognized, r and K selection, each of which is particularly adapted to characteristic patterns of change in the biotic or abiotic environment. As we shall see, there are problems with the oversimplification, but it is useful for the student to be aware of this distinction.

Two main modes of selection are postulated to occur, as suggested by the Verhulst-Pearl or logistic equation:

$$dN/dt = rN - (r/K)N^2$$

where r is the maximum rate of increase that a population of initial size (N) exhibits when released from exogenous limitations (i.e., not food limited, little predation, etc.), and K is the carrying capacity of the system, equivalent to the virgin biomass or carrying capacity of the environment in the usual fisheries formulation. Pianka (1972) notes that although some species are particularly adapted to rapidly occupying vacant spaces in the ecosystem (r strategists), others are better adapted to slowly increasing their share of the resources of the system under conditions of high competition (K strategists).

The r- strategists are species adapted to living in environments where there is a high density-independent mortality rate. Such areas include for example, upwelling zones, and those high latitude seas where (for annual species at least), seasonal periods of high food abundance are followed by periods of relative scarcity. These species tend to allocate a greater proportion of their resources to reproduction, having in consequence a high gonadosomatic ratio and birth rate, a short life span, as well as being in general relatively unspecialized, especially for example in feeding habits. Population size is generally prone to high fluctuations in numbers (e.g., Engraulid and Clupeid species).

By contrast, K strategists are adapted to living in environments where intra-and interspecific competition is high. These tend to be mature, relatively stable environments, such as coral reef and mangrove areas in the tropics. There is consequently a high degree of selection by means of density dependent factors, and these species have evolved to allocate a greater proportion of their resources to non-reproductive activities promoting individual survival. They are consequently relatively specialized in their trophic activities and other behaviours, and in general are longer-lived, less fecund, and less prone to major short-term changes in population size (e.g., sharks and whales: see Table 9 for a review).

Table 9

Some of the characteristics of r- and K- selected species (from Caddy, 1984)

Characteristics	r-selected	K-selected
Climate	Usually variable and/or unpredictable	Fairly constant and/or predictable (or species shows migratory behaviour)
Risk of natural death	Often high or catastrophic: largely independent of population size	Death rate is more scheduled and dependent on population size
Population size	Variable in time, non-equilibrium conditions prevail; occupies ecological vacuums but rarely reaches the carrying capacity of the environment; recolonizes habitat each year	Fairly constant in time, at or near carrying capacity of environment; no recolonization necessary of saturated communities
Competition between species and within species	Generally lax	Usually keen
Length of life	Short (usually one or less than one year)	Longer (usually more than one year)
Natural selection in favour of:	(1) Rapid development (2) High rate of population increase (3) High rate of egg production (4) Small body size (5) Single reproduction (6) Less emphasis on behavioural and morphological characteristics to increase individual survival survival habits	(1) Slow development (2) Low rate of population increase (3) Low rate of egg production (4) Large body size (5) Multiple reproduction (6) Behaviour and morphology assures good individual survival, e.g., territorial behaviour, spines, special dentition and special feeding habits
All above lead to:	Productivity	Efficiency

In his general review of the field, Pianka (1970) notes that in terrestrial systems there is a very clear distinction between the two strategies, and that in fact, for terrestrial animals this results in a clearly bimodal distribution of sizes, between small (predominantly annual) arthropod-dominated species, and larger, often perennial vertebrates; epitomizing in a rather clear-cut way the distinction between the two types of selection. He notes that to survive with a generation time longer than one year in a strongly fluctuating seasonal environment, a species must be adapted to the full range of ecological conditions encountered during a year. This is in contrast to annual species, which typically encounter only a limited range of seasonal conditions at any given life history stage, to which they can be perfectly adapted, and if necessary can pass the remainder of the year in an inactive "resting" stage, as eggs, spores, larval forms, etc. This suggests that generation times exceeding one year may be the threshold event in the evolution of a species, and will be characterized by a rather drastic shift from r to K selection modes. However, Pianka notes that fish show the full range of the r-K continuum, probably because (especially in tropical environments), seasonal changes in the ocean are strongly buffered, except for regions of environmental instability (e.g., upwellings, intertidal and estuarine regions) where it may be expected that r selected species are dominant. A summary of the main characteristics of r and K species is given in Table 9.

More recent work has cast some doubts both on this duality, and on the logistic assumptions that underlie it: for example, the concept of an equilibrium value for the carrying capacity K of an ecosystem requires some careful consideration: the known variations in carrying capacity K in a given area, with time, causes one to wonder whether in managing such resources, one should adopt the usual steady-state equilibrium approach, the open-system non-equilibrium approach, or perhaps some combination of the two (Caddy, 1984). The answer to this question, though far from clear-cut, depends on the time scale being discussed: even "stable" populations, over a long enough time scale, show departures from average landings as calculated over the medium term.

In the fisheries literature, the steady-state approach is the one most often assumed to apply, however, and because the different levels of harvesting can be quantified, this allows one the opportunity to test the relative effect of different perturbations on the system being examined. It is primarily due to this useful characteristic that approaches based on the equilibrium assumption have dominated the literature. However, a description of the open system, non-equilibrium approach (Johnson, 1981) shows that it can be equally valid to consider ecosystem dynamics as a series of perturbations, rather than as a system in equilibrium, and the application of this alternative hypothesis in fisheries is discussed in Caddy (1984) and followed up by e.g., Allen and McGlade, 1986[1]. The logical differences are fundamental, but this duality in approach is a common denominator of nearly all fields of science which seek statistical predictions of likely future states of a system, e.g., in thermodynamics, small particle physics, the functioning of nervous systems, etc.

There are numerous examples of rapid changes in state occurring in marine biology: the abrupt transition of larval forms seen in many marine organisms at various stages in development is one example. Most fish start life as an independent egg adrift in an uncertain environment with few adaptations other than physiological for survival; therefore, their common "objective", expressed anthropomorphically, is to become independent of local environmental limitations by becoming mobile. This requires that they proceed as rapidly as possible through the developmental stages leading to increased mobility. This is classic r selection. Some other fish bear live young, which are already quite mobile at birth: a classic K selected process. However, fish that grow to large sizes from small eggs often become characteristically K strategists to the point of expending far more of their energy in activity than in reproduction. The oceanic nomad species are good examples (e.g., tunas, dolphin fish, billfishes).

Clearly there is no set of either-or conditions in regard to r and K labelling of fishes. This makes it quite difficult to quantify the utility of these concepts in application to practical ecological problems. However, there are interesting relationships between the intrinsic rates of increase, r, and population density. For example, looking a little further into the implications of the mathematical theory mentioned earlier, if we plot the rate of increase in population against population density as in Figure 40, we see that to the left of some density, X, the r-strategist is at a competitive advantage, while to the right, the K-strategist is favoured. It is interesting to note here that the plot of rate of increase against density in Figure 47A is functionally equivalent to a stock-recruit model, and further, although this analogy should not be carried too far, than there is a resemblance between the curve for r-strategists and the so-called Ricker Spawner-Recruit (S-R) curve, Figure 47B while that for K-strategists resembles the so-called Beverton and Holt S-R curve (Ricker, 1975). An example of the former in the marine sphere is the anchoveta, typically a short-lived opportunistic species, and of the latter, the North Sea herring. This latter is a relatively long-lived species, formerly occurring in great abundance, but with a demonstrated vulnerability to density-independent mortality effects. These latter we may consider to include human predation by fishing, especially on spawning grounds. In herring, fishing effort remained high despite declining abundance, in part because the species is vulnerable since it

---

[1] See also Sharp, 1986; Neritic systems and Fisheries: their perturbations, natural and man induced. In: Ecosystems of the World Part 28 Ecosystems of Continental Shelves (H. Postma and J.J. Zijlstraa, eds.) Elseviers Scientific Publishing Company, Amsterdam-Oxford-New York. 1986

occurs at a uniformly high density on the spawning grounds, whatever the stock size. The relationship between the form of the hypothetical spawner-recruit curve and the concept of r and K selection is logical, since both deal with the rate at which a stock replenishes itself under different densities. Obviously, with no parental stock, there will be no reproduction, but for most species it is still far from obvious what is the relationship between parental stock size and the numbers of progeny produced. In the S-R relationships postulated to date, the number of recruits drops off with biomass or density since as is inevitable in a finite habitat, there is only limited space and food for a certain number of individuals. Line A-B in Figure 47B would only then be approximated in a continually expanding environment. Expressed somewhat differently, although we should ideally try to describe what is the relative contribution to reproductive success of parent stock size, environment, and interaction of recruits with other organisms (predators or competitors) in the ecosystem, this will be no mean task, and presupposes first that we have a good understanding of the life histories of the key component, and their interrelationships.

Reviewing the field of density-dependent recruitment is unnecessary here, but a few of the underlying concepts from this extensive literature (see Ricker, 1975 for an introduction), need to be placed in an ecological context.

The overall carrying capacity K of an ecosystem can only be defined in the context of its basic productivity, and the relative efficiencies of population growth and food conversion, and hence abundances, of the component species within it. The thermodynamics of energy acquisition and conservation is the fundemental 'preoccupation' of species in their life history strategies, but this is expressed through the activities of work-growth-reproduction and relocation, so that the fundamental problem for an ecologist is of defining, measuring and monitoring these processes. To assume for a species or stock a simple relationship, where recruitment is only a function of parent stock size, is to start well above the relevant levels of interaction, and is unlikely to yield satisfactory insights into the dynamics of either ecosystems, or even of one component population.

The concept of the ecological niche in the marine environment

In many, if not most oviparous species, the habitat for larval survival is so specific that the carrying capacity of that subset of the species' overall habitat or 'niche' may well be the main constraint on the species in question. Salmonids (Ricker, 1975) and engraulids (Lasker, 1978, Bakun and Parrish, 1980, Sharp, 1980b, IOC Workshop Report No. 28), appear to exemplify these constraints in action.

The following enlightening discussion of the likely impact of multiple and overlapping ecological niches for various marine fish species, on our perception of "competition" for food, is taken from Jones and Henderson (1980):

"The possibility of larval niches of different sizes, as well as adult niches of different sizes raises other considerations in connection with the full utilization of food resources. Thus if the population size of a particular species is regulated prior to the adult stage it may appear possible to have a non-limiting food resource for the adults. This however, must be a short term viewpoint. In the long term, other species may be expected to take advantage of any surplus food, and eventually all resources should be fully utilized. The problem then is to explain how it is possible for the adult population of any one species, to expand suddenly, due to the influence of one or more good year-classes."

An understanding of how this might be possible comes from a consideration of larval and adult habitats as separate; each with its own competitors and interspecific competition. Thus, larvae or juveniles of a species exploit a particular resource, and the adults exploit a different one, consisting of larger-sized food particles. If the adult resource is not fully utilized, there is room for the adults of a second species (Y) to share the adult resource. However, it should be possible for both species to co-exist providing their larvae exploit different resources. This process of adding species can be expected to continue until all the adult food resource is completely utilized, and all larval niches are occupied. (We may note that the addition of "new" species in this way is occurring continuously by evolutionary processes).

Continuing, Jones and Henderson (1980) note that "In a multispecies situation of this kind, ... total recruitment is less variable than the recruitment of individual species. A temporary expansion of one species could then be offset by a temporary decrease of other species. This is an attractive argument since it provides an explanation of how it is possible for the diets of many fish species to overlap, without competition reaching a level at which one or more species is eliminated. It also explains how it is possible for the adult stock of any one species to increase significantly, due to one or more good yearclasses, without exhibiting evidence of significant resource limitation. Resource limitation of adults will depend on the combined biomass of all species sharing the same resource, and this may vary very much less than the stock sizes of the separate species. The assumption that there is an upper limit to recruitment due to resource

- 76 -

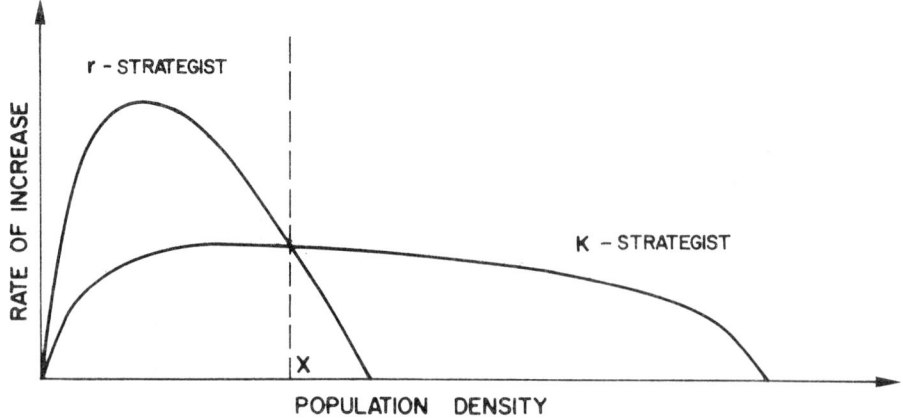

Figure 47A Postulated relationship between population size and rate of population increase for r- and for K-strategists

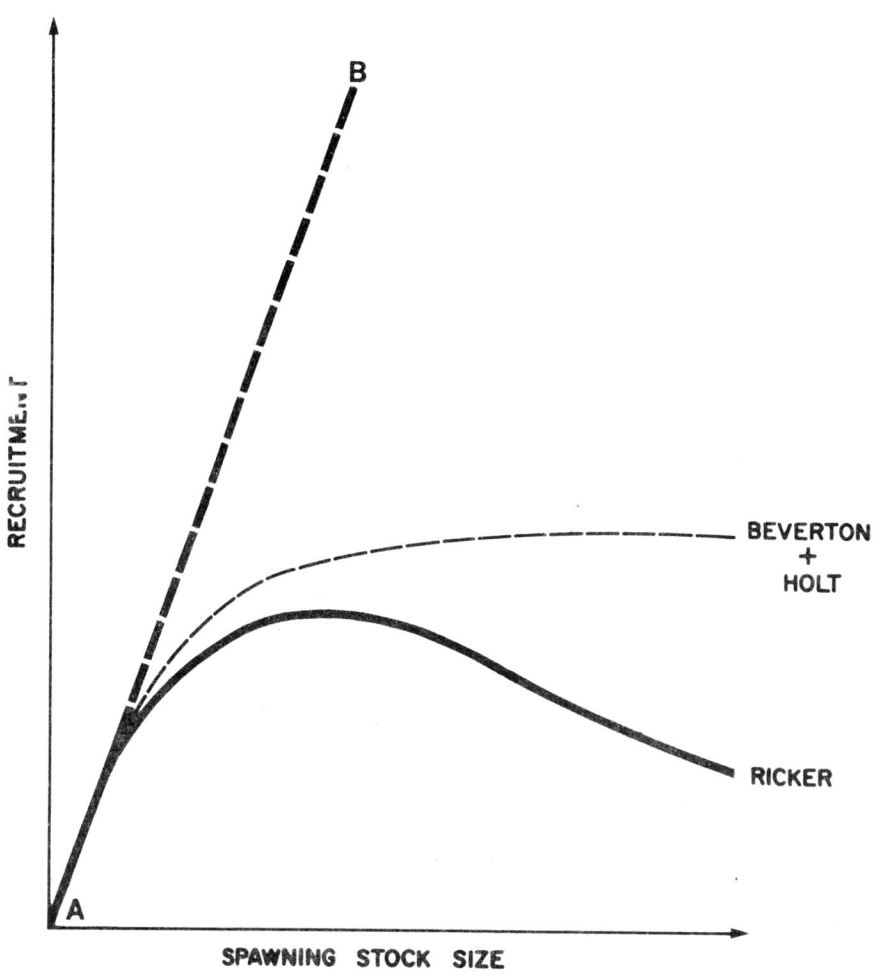

Figure 47B Two relationships between spawning stock size and subsequent recruitment in common use. (Line A-B is the trajectory that would apply if recruitment were proportional to stock size.) See Ricker (1975) for details

limitation during the early life is attractive because it makes it possible to understand how a relatively large numbers of species might co-exist. Given differences in time, as well as in space, the potential number of larval ecological niches must be very much larger than the potential number of adult ecological niches."

Given that the absolute number of fertile eggs produced puts a ceiling on the possible size of a new generation, the question of where and when they should be laid seems the next logical question to ask. Homing species and localized populations are both tuned by selection to "fit" the occupied habitat, or become extinct. The significance of this statement is often forgotten until a massive change is observed in the range, distribution and abundance of a species or groups of species. Recently Parrish and MacCall (1978), Csirke (1980), Grainger (1980), Iles and Sinclair (1982), Sharp (1980a; 1980b), Bakun and Parrish (1980) and Sharp and Csirke (1983) have documented and tried to elucidate the time-space and ecological implications of physical and biological variations in the environment and their effects on population level variations and recruitment success, as well as on distribution and abundance. The problems of time and distance scales need resolving before useful questions can be answered (Sharp, 1980a; Fasham, 1978).

In general, the perspective emerging from even this brief discussion, is that for many marine species with multi-stage life histories, the idea of an ecological niche as the sum of all physical and physiological influences on the species, becomes virtually impossible to define in quantitative terms. It is perhaps also incorrect axiomatically in its implication that each species is unique in its position in the ecosystem: a growing realization of the wide degree of overlapping in life history stages and their trophic preferences is perhaps a better description of current thinking in this area.

A series of questions, and a brief illuminating discussion of the 'multiple' nature of niches in the marine ecosystem is extracted from Jones and Henderson (1980):

"For a long time, fishery biologists have been speculating about the dynamics of fish recruitment and the method of population regulation in teleost fishes. Associated with this problem are important management questions, such as "how hard can a particular stock be exploited before there is a risk of recruitment failure and a collapse of the fishery?"

There are also important ecological questions such as:

"How are fish populations regulated, given that there is apparently no correlation between stock size and subsequent recruitment?"

"At what stage in the life history is the level of recruitment determined?"

"How is it possible for a good yearclass, to lead to a significant increase in adult stock size, without the individuals in that stock showing obvious signs of resource limitation?"

"How is it possible for so many species (particularly in tropical waters) to co-exist, when there is apparently such a large overlap in their diets?"

"In the present state of knowledge, it is not possible to arrive at defnite answers to any of these questions. All that can be done, is to provide a framework, within which plausible answers to these questions suggest themselves."

In relation to the definition of recruitment level and stock size, Jones and Henderson (1980) note that: "Since different life stages exploit different food resources, it becomes important to distinguish between the average level of recruitment, and the average size of the stock. Thus, average recruitment (R) each year, should be largely determined by the size of the juvenile food resource but, average larval production (L) each year is likely to be determined by the size of the larval food resource.

Adult stock size (A) on the other hand is more likely to depend (in the long term) on the size of the adult "ecological niche" than on the number of recruits produced annually." In relation to the questions of resource rarity and dominance they also note that:

"Where there are many species, it seems likely that those that succeed in exploiting large resources as adults, will also succeed in exploiting large resources as larvae and juveniles.

In this context, a large resource might not only mean a large geographical area, as suggested by Iles and Sinclair (1982) but also an extended spawning period, or both."

r- K-selection theory and stock recruitment

Garcia (1982) has documented the illogicality of assuming a single stock-recruitment relationship for penaeid shrimps, a group where the adult and earlier stages are ecologically separated, hence few direct intraspecific effects beyond egg production can reasonably be assumed. This example shows in the simplest case of an annual (r-strategist) that the environmental conditions during spawning and recruitment can be at least as important as stock size in determining the size of the next generation. There are clearly also other parameters which may determine the realization of recruitment potential, by affecting processes such as transport, predation, etc., and which modify the potential represented by the viable, fertile eggs produced by the spawning stock.

The limited utility of the single species spawner-recruit paradigm becomes even more apparent when one tries to describe the effects of one species' failure to take advantage of a series of ecological blooms in its prey species, or in the prey of its larval form. This failure may perhaps be due to low predator abundance, hence to a poor ability to take advantage of opportunities. These "missed opportunities" may well (in fact, are likely to), be taken up by another (competitor) species, yielding a bloom in this normally subdominant species which in this context can be regarded as an r-strategist. Perhaps the sequences of anchovy-sardine blooms evident in the literature on pelagic fishes represent just such scenarios (Soutar and Isaacs, 1974). Skud (1982) has focussed his attention on the impact of such changes in species dominance in modifying the correlation between ambient conditions and the population characteristics as sampled by fisheries, of the species that alternate in dominance.

From analysis of fish otoliths in bottom sediment cores, it seems that such species changeovers were occurring regularly in areas where high levels of pelagic production are now the case, for tens of thousands of years prior to human intervention in the form of a fishery on the species components (Devries and Pearcy, 1982). This raises the question, not yet adequately answered, of the relative importance of high fishing intensity and environmental instability in determining population changes, particularly for species such as the Peruvian Anchovy (The Costa Rica Symposium: Sharp and Csirke, 1983).

r-K selection theory and natural mortality rate, M

A further relationship between r and K theory and conventional fisheries dynamics, concerns the natural mortality rate, which may be considered as being inversely related to the individual fitness of an organism to survive.

A study of a limited number of (unfortunately, largely arctic-boreal) fish species (Gunderson, 1980), supports the prediction from r-K selection theory, that natural mortality should increase as one moves across the spectrum from K-strategists to r-strategists.

It was noted that there have already been a number of attempts to estimate this very difficult and important parameter (M) indirectly from its correlations with various other more easily estimated parameters. Thus, mainly for terrestrial organisms, a strong positive (nearly linear) correlation exists between the logarithm of body length and the logarithm of generation time (i.e., in conventional fisheries terminology,

$$\log (L_{oo}) = a + b \log (1/M)$$

but Gunderson (1981) found body length to be the weakest variable for predicting M of the four variables he used, the others being:

1. Age at 50 percent female maturity (T_m);

2. longevity (the age $T_{.01}$ by which abundance has declined to one percent of the average numbers recruited under unexploited conditions);

3. The von Bertalanffy growth rate, K;

4. Gonad index (GI) defined in its simplest form as gonad weight for mature females. (This index is better than body weight as a measure of fecundity, since it takes into account both spawner population and number of eggs produced per female).

This last measure (GI) proved to be the single best correlate with M of those considered, so that the following regression was highly significant:

$$M = 4.64 \text{ GI} - 0.370$$

The addition of the second most useful variable ($T_{.01}$) in a multiple regression to predict M, only decreased the standard error of the estimate by a further 12 percent.

As noted by Beverton and Holt (1959) and Southwood (1976), the von Bertalanffy growth constant K' for individual growth (not identical with the Verhulst-Pearl K!) is a good correlate of M within taxonomic groups, as would be expected from r-K selection theory, since short-lived animals generally have high metabolic rates and individual growth rates; once again well correlated with K. However, correlation between the two variables is rather poor when members of more than one taxonomic family are included. Nonetheless, Pauly (1980) using a multiple regression approach similar to Gunderson's, found that a first estimate of M could be gained from a knowledge of K, Loo and mean environmental temperature ($\bar{T}$), and suggested formulae for obtaining M from a knowledge of these. His multiple regression equation was given earlier.

Definition of $\bar{T}$ for some of the larger, especially the mobile species in Pauly's list is quite arbitrary however, and the wide variance around the relationship must encompass not only a significant measurement error, but also variability due to fluctuations in predator-prey relative abundance and prey availability which are not included within the three independent variables. It is therefore somewhat surprising how well this equation succeeds in establishing an "order of magnitude" for the population mortality rate.

Gunderson's equation using GI cannot be used as given for tropical species, both because it would extrapolate beyond his very limited sample (very few of the relevant data sets are available for tropical species), and also because (multiple) spawning of tropical species often occurs over a longer period of the year than for temperate species, so that the Gonad Index would have to be adjusted for repetitive spawning to be a good index of either M, or the species position on the r-K spectrum. This criticism also invalidates the equation of Rikhter and Efenov (1976) for use in tropical waters. A comparison of known M's with values of E, would be needed, (where E = number of eggs shed per mature female) x f_o: f_o is the mean number of spawnings (including partial spawnings) per year by mature individuals. This approach would be of interest, but for the moment the necessary data sets are not available to test it.

r- and K-selection and the "ecological niche"

The concepts of species interaction and competition are also easily placed in the context of r- and K-selection theory: competition for a limited supply (of space, food, etc.) will inevitably increase with crowding, so that one can predict that the maximum tolerable niche overlap will be greater in relatively saturated communities (i.e., K-selection conditions), and one may expect more exotic adaptations, behaviourally and in terms of body structure, to occur as species diversity increases.

As noted earlier, Baisre (1985) described in sequence three Cuban ecosystems (estuarine, reef-associated and oceanic), in terms of the decreasing influence of land [Figure 48(A-C)]. With distance offshore, dissolved nutrient levels diminish, "particles" (organisms) increase in size, as does the index of diversity and the maturity of the communities, although Baisre notes that some organisms may move offshore along these gradients in the course of their life histories. Concomitantly, he speculates that a move from r- to K- life cycle strategies occurs as unit productivity decreases. An even more sweeping generalization, which nonetheless has a degree of truth, is that while spawning and hatching tends to occur for many species in stable environments such as eel grass beds (see Pollard, 1984) where natural mortality is reduced, larval and/or juvenile life histories may take place in more unstable environments where the food supply is greater.

Judged from the preceding account, r-selected species would seem less manageable than K-selected ones. This conclusion must be regarded with caution, however, because such a simple two-way classification does not take into account that we may be talking about a function of environmental instability as opposed to just a function of life history strategies. Caddy (1984) notes that "it seems quite possible for a short-lived r-selected species to achieve and maintain dominance in a stable environment if the population of K-scheduled competitor species is kept low by fishing." Some cephalopod and shrimp populations seem to illustrate this, even if longer-lived finfish components have been cropped down.

An alternative classification to r-K theory has been proposed by Kawasaki (1980) that leans heavily on Japanese experience in N.W. Pacific Ocean. This is shown in Table 10, where a long-term history of fluctuations in production tied to the Kuroshio Current system provides an interesting perspective to the opportunistic and varied approaches to a wide resource basis adopted by the Japanese fishing industry.

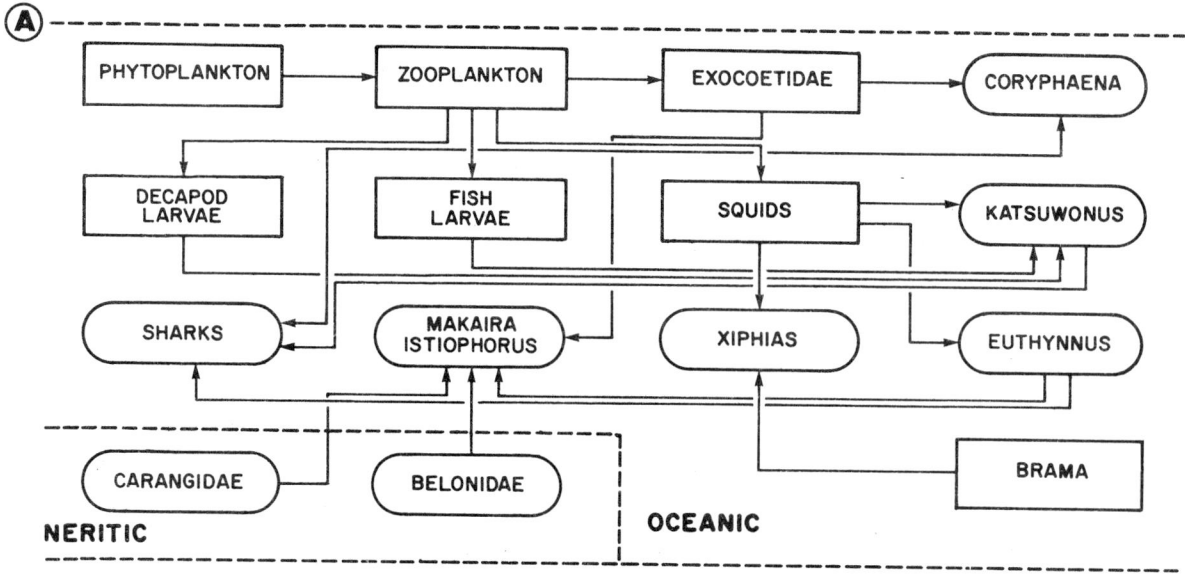

Figure 48A General scheme of feeding relationships in the oceanic water complex. Data from various authors cited by Baisre (1985)

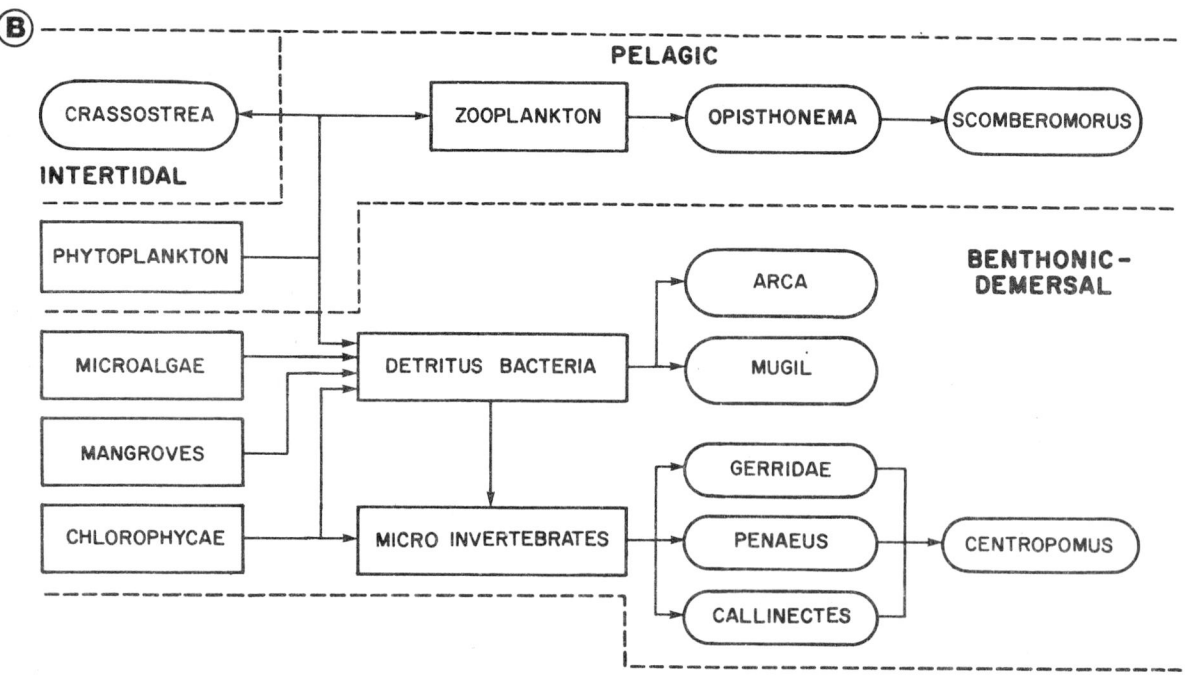

Figure 48B General scheme of feeding relationships in the littoral estuarine complex. The names in circles correspond to species or commercial groups. Information from various authors given by Baisre (1985)

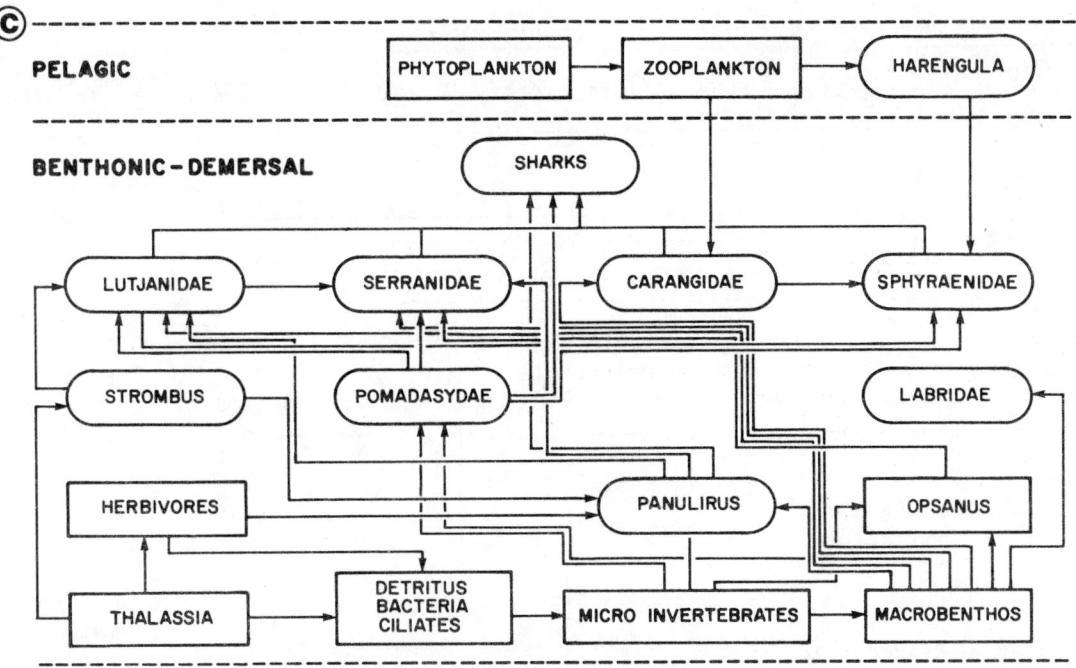

Figure 48C   General scheme of feeding interrelationships in the ecological complex of reefs and turtle grass beds. Data from various authors cited by Baisre (1985)

In Kawasaki's scheme, two main and two subsidiary classifications of species are proposed: Type II being similar to the K-selected category above, except that number of eggs produced per spawning is high and includes, as examples, cods and flatfish (e.g., Type 4). Being generally high in the food chain as adults, these species may be considered manageable by conventional yield models, with equilibrium concepts being largely applicable.

Type I is divided by Kawasaki (1980) into:

Type IA - small short-lived species with low individual fecundity but very early maturity, early spawning and hence high rates of population increase. These species (e.g., saury, squid and sand eel) tend to inhabit irregular variable environments and show high population variations, making them essentially very similar to the r-classification prescribed earlier (except for their low fecundity), and they may be considered non-equilibrium stocks where pulse fishing may be the only way of harvesting the stock without excessive wastage due to natural deaths in good years, but where effort will have to be directed elsewhere in poor years. Here the manager is faced with the conflicting considerations of deciding between a high enough effort level that reduces wastage and a high probability that alternative fleet deployment to other resources will be needed if population crashes occur, as usually seems to occur at some point in time with this type of stock (e.g., Figure 18).

Type IB - Here we have stocks adapted to areas with long-period fluctuations or cycles in suitability of the environment for stock recruitment within the cycles. This group is somewhat intermediate between the other two in that life spans are long enough, but age at maturity is generally early, with multiple spawnings and moderate fecundity allowing a rather high rate of build-up of population biomass over the medium-term (5-10 years?) period of favourable conditions, and with sufficiently long life span for some individuals to survive to the next favourable period (e.g., sardines, herring).

Table 10

Correlates of three selected tyes of life history in marine teleosts (after Kawasaki, 1980)

	Type I			Type II
	Sub-Type A		Sub-Type B	
Environment	Irregular variation	Variable and unpredictable	Long period variation	Stable and predictable
Recruitment	Irregular variation	Variable	Long period variation	Stable
Resources put into:	Reproduction only	Reproduction	Reproduction % maintenance	Growth and maintenance
Lifespan:	Short		Long	Long
Growth:				
Growth rate	Moderate		High	Low
Maximum size	Small		Moderate	Large
Reproduction				
Age at first maturity,	Very low	Low	Quite low	High
Fecundity	Low		Moderate	High
Intrinsic rate of population increase	Very high	High	High	Low
Early survival		Variable		State
Trophic level		Low		High
Typical species	Saury, sandeel		Sardines, herrings	Cods, flatfishes
Appropriate management measures	Catch forecasting and monitoring (+ pulse fishing?)		Recruit forecasting & yield/recruit assessment & MSY fishing?	Equilibrium yield assessment, steady state ($F_{0.1}$) fishing.

11. FOOD WEB ANALYSIS IN PRODUCTIVE COASTAL ENVIRONMENTS: THE COASTAL KELP, SEA URCHIN AND LOBSTER SYSTEM IN HIGH LATITUDES, AND SEAGRASS NURSERY AREAS IN THE TROPICS

The interesting question which it is obviously most difficult to answer from short term research, is the extent to which an ecological system or community is subject to changes or fluctuations over the long term. Perhaps one of the first systems where repeated "switching" between two dominant ecosystems has been studied in some detail is that referred to as the "Russell cycle" (see Cushing 1982 for a description and references). Briefly, a change in ocean circulation causes major changes in characteristics of water masses, plankton and dominant pelagic fish stocks in the English Channel and Southern North Sea: principally between dominance of pilchard and herring stocks. Similar events have occurred elsewhere in the world, for example, off the western coast of South America, and off Japan and California (see papers in Sharp and Csirke, 1983), to mention only a few areas where dominant pelagic stocks have shown significant changes in species, biomass and distribution.

Such events also occur in the inshore environment and the benthos, and one such example from high latitudes is selected here because it illustrates the dangers of assuming (a) stability, (b) irreversibility of marine ecosystems, and also some of the difficulties ecologists face in trying to evaluate the mechanisms of change even in relatively accessible near-shore systems.

Figure 49 from Miller, Mann and Scarratt (1971) used the energy circuit language of Odum to show the results of an early analysis of energy flow within the sublittoral kelp (seaweed) community off Southwest Nova Scotia, in which in situ measurements of biomass per unit area, and estimates of respiration rate were used to quantify relative rates of transfer of materials between components of the ecosystem; each expressed in terms of the production per square metre of the sublittoral zone, using the relationship between rates of respiration and production derived by McNeill and Lawton (1970).

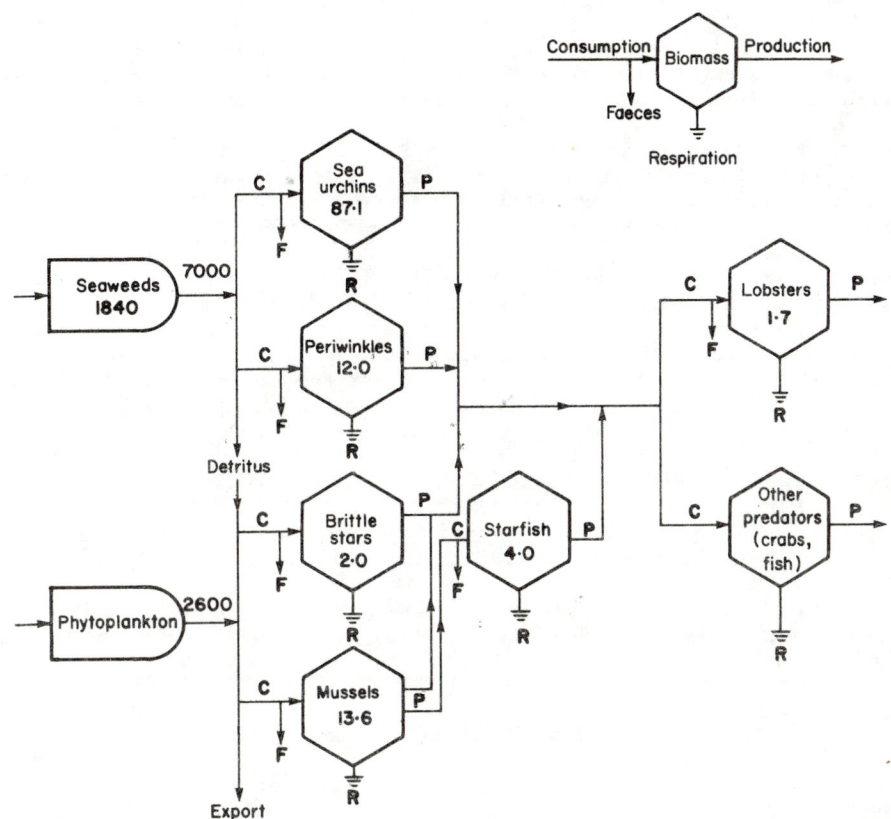

Figure 49   Energy flow through the sub-tidal benthic community in the seaweed zone of Saint Margaret's Bay, Nova Scotia. Units are Kcal/m^2/year, except for biomass which is in Kcal/m^2. Phytoplankton production figures are from Platt (1971) (After Miller, Mann and Scarratt, 1971)

Explanation to Figure 49: Figure 49 uses the energy-circuit language developed by H.T. Odum (see Platt, Mann and Ulanowicz, 1981 for details), where the seaweed bed is viewed as a combination of energy receptor (through photosynthesis) and self-generating consumer unit (by respiration): with a standing stock of 1840 Kcal/m^2/yr, which contributes 7000 Kcal/m^2/year to the population of herbivores and detritus feeders supported by it (each in turn representing a self-maintaining consumer population).

As shown in Figure 49, consumption of plant material produces faeces and detritus supporting benthos, and also shows a very significant net loss in the chemical energy content of food web constituents by respiration, as signified by the symbol indicating return to ground; (or to "sink" in the more usual energy circuit language).

The first important conclusion from this calculation is that annual seaweed production exceeds herbivore consumption by more than ten-fold: much of this material which is not consumed in situ, plus "secondary" detritus in the form of faecal material is exported from the local subsystem as broken pieces of fronds and other plant debris, and like other sources of detritus elsewhere, e.g., mangroves and seagrass beds, makes an important contribution to the food chains of adjacent shelf regions. This contribution has been estimated by Mann (1972) to be approximately equal to that from phytoplankton within the 90 m depth contour off Nova Scotia: illustrating its potential as a source of organic material for near-shore food webs.

Secondly, lobsters, starfish and other primary predators such as the wolf fish Anarhichas lupus control the population of herbivores, of which the most important is the sea urchin, Strongylocentrotus droehbachiensis, which is a major grazer of algae, and which in the 1960s showed major increases in abundance associated with a serious decline in kelp stocks. Absence of control of urchin populations due to overharvesting of primary predators was offered as the main reason for serious depletion of kelp beds in the 1970s from many apparently suitable rocky substrates along the Nova Scotia coast by Breen and Mann (1976) and Chapman (1981). In the 1970's it was also clear (Pringle et al., 1982), that overfishing as well as shortage of suitable habitat at key "bottlenecks" in the life history, rather than just limitations on food, were likely to be the responsible factors for the low population size of lobsters, and given the wide range of food accepted by lobsters, overfishing or loss of shelter may have been the most important effects.

In the light of extensive subsequent discussions held on the nature of the interaction described in the paper by Miller, Mann and Scarrat (1971), (see Pringle et al., 1982 for a review), it seems clear that although interesting, it is not evident here what the measurement of the energy flow in one or two limited locations has contributed to the resolution of the relative importance of the linkages described. This is partly because several essential pieces of information were missing at that time, related to the relative size and spatial interractions of the individual species components of the system, and not just their rates of feeding. It also is becoming clear that components not included in the preliminary energy-circuit diagram above may play an active part. This is also the case in the similar sea urchin - Macrocystis - Abalone - sea urchin - sea otter system referred to in Pringle et al. (1982), where the key role of other components (a sea star and labrid fish, not to mention the possible effects of coastal pollution), have all been considered as important factors in producing the community changes observed, in addition to the previously identified components mentioned above. It is clearly important to have some idea of the relative feeding rates and spatial overlaps of a given predator on one or other prey species.

The aspect that most occupied workers on this Nova Scotian system in the 1970s was the apparent irreversibility of the increase in urchin abundance, and the decline in kelp biomass, but as will be seen in the following, it is important to also be aware of the broader zoogeographical scale and longer term history of a system before drawing rapid conclusions. The above change to urchin dominance appeared irreversible because following active grazing down of mature kelp stands, urchins, although showing stunted growth, were able to persist and keep 'barrens' free of kelp sporlings while grazing on detritus material and other plant sporlings settling on the rock surface. At this point, Breen and Mann (1976) concluded from all available evidence, that it was "difficult to visualize a mechanism that would allow kelp beds to reappear".

In 1980, such a mechanism made itself felt dramatically in the form of a sea urchin pathogen (Miller, 1985, 1985a), rather similar in its impact to that which in 1983 decimated populations of the tropical urchin Diadema antillarum throughout large areas of the Caribbean (Lessios, Glynn and Cubit, 1984). The Nova Scotian disease vector proved virulent and easily transmitted, and in two years (Miller, 1985, 1985a), some 250 000 t of sea urchins are estimated to have been killed. Kelp recovery rapidly followed, and the area of habitat now released to kelp production is estimated potentially to support a seaweed crop of some 1.8 million tons, and an annual production of some 7 million tons. It will be interesting to see what other ecological effects follows from this greatly increased production of coastal organic material; not only on the lobster stock, but on coastal production in general.

As a result of this experience, circumstantial evidence was collected that such urchin epidemics had occurred in the past, and Miller (1985) postulates a cyclical succession of fairly long period (Figure 50), with both urchin and kelp alternating in dominance.

The picture now seems a little clearer and more optimistic than in the 1970s for this system, though as in most ecosystem research, as many new questions are raised as were answered. For example, if the pathogen is so virulent, where did it come from? What will be the relative contribution of increased habitat versus food to coastal resources such as lobster that occupy the kelp zone? The one fact that cannot be doubted, is that ecosystem stability is less certain an axiom now than it seemed to be as little as 10 years ago.

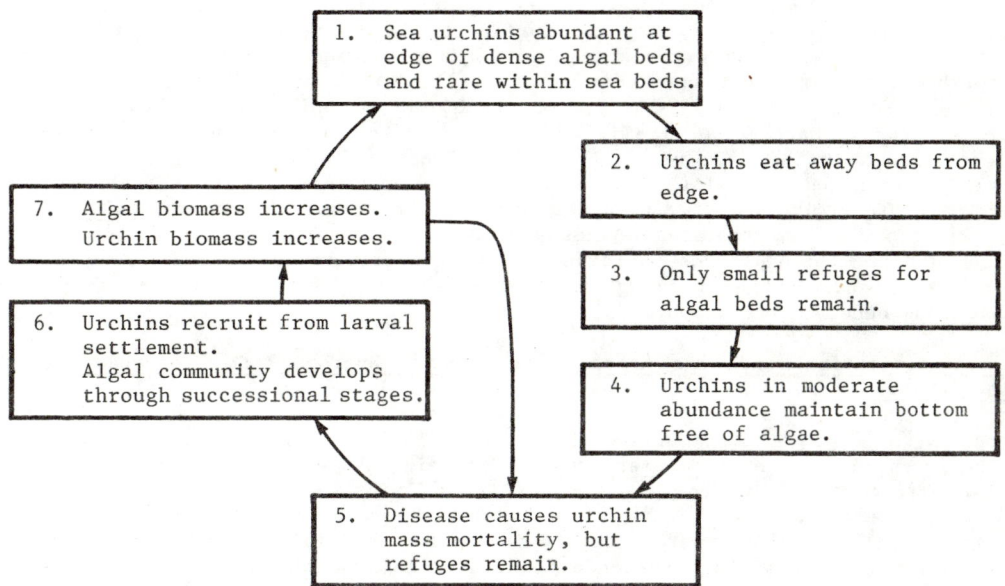

Figure 50   Seaweed-sea urchin cycle (from Miller, 1985a)

Seagrass nursery areas - juvenile and feeding habitats for many reef-dwelling species

Seagrass beds occur over a wide range of latitudes, from Tasmania to Japan and the Baltic Sea, (see review by Pollard, 1984), and in all areas play an important ecological role, which seems only in part to relate to their importance as a contributor to detritus food webs, especially leading to forage species of small epibenthic crustacea such as amphipods. These latter are important in the diets of early stages of juvenile fish; many of which use seagrass beds as shelter from predators. This habitat thus provides shelter to juveniles of fish species and lobsters, and is a foraging area for predatory and herbivorous species, especially in the tropics from adjacent coral reef areas. These feed over seagrass beds, often on a nocturnal cycle: returning to the reef areas during the day. This observation suggests care should be taken in assuming that the very high production estimates observed in some coral reef areas are entirely due to primary production from reef areas alone.

There seems to be some question in the literature as to how much of the seagrass production results directly from its grazing by herbivores: like other macroscopic plants we have mentioned, much of this plant production seems to enter the food web through detritus: especially through the crustacean linkages to juvenile fish. The high figures for fish production of seagrass beds apparently results from the large number of young of the year which are growing at very high rates (Thayer, Adams and La Croix, 1975).

One other feature of this and other types of marine vegetation (as well as other types of outcrops from the sea floor), is the high surface area they provide for larvae settlement, and for epifauna and flora: this perhaps is of comparable importance ecologically to the direct contribution of seagrasses to herbivore food chains, and this aspect of extensive substrate surface and shelter from predators must be a common feature of many fish nursery areas. In the case of seagrass, it may be that a change in the fractal nature of the habitat makes it a less secure shelter for organisms over a certain size: the size at which many juvenile fish move to other habitats. Also, despite their importance at key juvenile stages, it has been noted that grass beds, with some exceptions, are relatively less important as fish spawning areas.

The growing understanding of the significance of such centres of primary production in the nearshore and littoral zone, and the potential impacts of a variety of human activities (e.g., dredging, mine runoffs, urban wastes, etc.) on these habitats, constitutes one area in which an application of the constraints of marine ecology contributes directly to maintenance of healthy levels of near-shore fish production.

Experimental work is now beginning to accumulate on the possibility of restoring depleted seagrass beds (e.g., Thorhaug, 1985), following large scale interventions in near shore ecosystems.

12. FOOD WEBS IN LOW AND HIGH ENERGY SYSTEMS: A MEDITERRANEAN DEMERSAL FISH COMMUNITY VERSUS FISHERIES OF UPWELLING SYSTEMS

An example of the construction and potential use of qualitative food web concepts in fisheries management is the demersal fish community of the shelf and slope of the Baleares and Western Mediterranean: a fauna that is intermediate in complexity between relatively species-poor areas such as the northwest Atlantic and upwelling areas of the world, and the much more complex and diverse fish communities of tropical areas. A matrix of the trophic inter-relationships of this fish community is shown in Macpherson (1981), who also describes some of the important benthic and planktonic invertebrates in the food web. These lower components of the three "trophic levels" shown in Figure 51 are given an arbitrary trophic number, beginning with those fish species - the inner circle - which are exclusively invertebrate feeders. Interestingly enough, Molva is the only fish species here in which invertebrates do not play any significant role in the diet. Although they are not shown here, it should not be forgotten that invertebrates themselves do not occupy a single trophic level, and that even among invertebrate feeding fish, a considerable degree of resource partitioning occurs (MacPherson, 1981).

Trophic classification and the compartment model approach to representing trophic interrelationships

One alternative approach that summarizes the whole of Table 1 in Macpherson (1981) is shown in Figure 52, which is essentially a compartment model indicating where the energy of the system comes from. This is instructive, from the structural point of view, in that two separate groups of "forage fish" may be recognized in the two large basal circles, largely feeding on planktonic and benthic food respectively; (although here a degree of overlap certainly occurs: nearly all benthic feeders must occasionally take some planktonic food, espeically early in their life history, and possibly to a lesser extent, vice versa).

Considering the higher level predators, Figure 51 underlines the fact that for most of them, invertebrates still make up a significant part of their diet. Despite this, three basic predatory groups may be recognized:

1. Predators on largely planktivorous fish (Micromes and Scyliorhinus);

2. feeders on largely benthophagous fish (Phycis); and

3. mixed feeders, which consume most species.

From a research perspective, and bearing in mind what we have discussed earlier, some possible implications of this distinction that seem to merit further investigation are:

(a) population sizes of category 1 species and their prey are a priori likely to be more volatile than category 2 species;

(b) categories (1) or (2) species are less likely to be reciprocally affected by management measures on species in the other category, than species in the same category.

Other than this, we may consider compartment models as biologically interesting and useful in demonstrating how biomass or energy is partitioned, but if based on purely trophic criteria, of less immediate application to management of fisheries systems. Consequently we concentrate in this account on Figure 51) from now on, which describes individual species interractions.

The use of a circular format for the food web in Figures 51 and 53 is largely a graphical convention which stems from the need to maintain intelligibility by at least ensuring that crossover of linkages between components is minimized, and when it occurs, remains at as close to right angles as possible. Conceivably, it would have been more logical to have reserved the inner

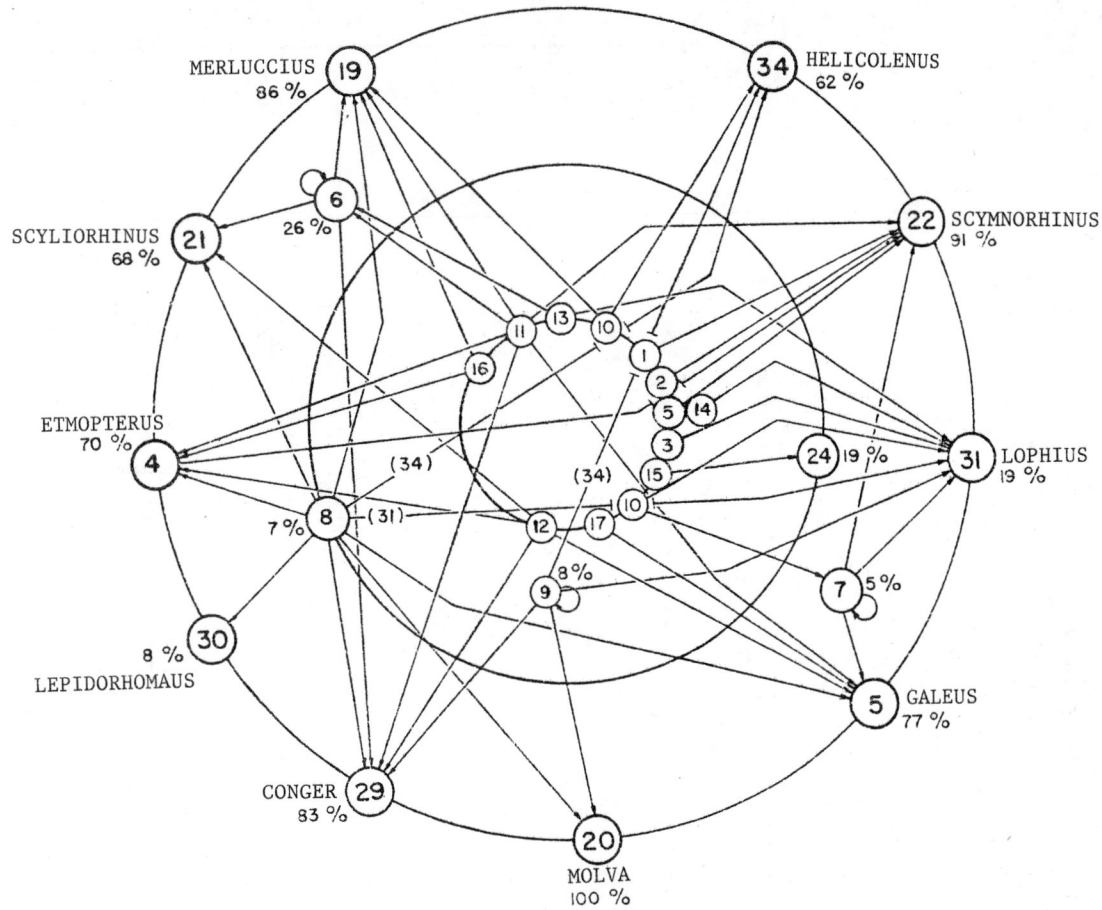

Figure 51   Showing detailed food web linkages for a Mediterranean demersal finfish community after MacPherson (1981). The "apical" predators are shown in the outer of the three concentric circles, and "forage" fish (purely invertebrate feeders) in the inner circles. Percent points by each component indicate percent of fish in the diet. Components (e.g., Nos. 6, 7 and 9), with self-directed arrows, are in part cannibalistic

circle for the apical predators (see, for example, Figure 53), but since the commercial emphasis is on those economically more important species generally higher in the food web, which are here shown with larger circles, the present arrangement has some underlying logic, as well as conveying the basic idea of dissipation of benthic productivity upwards through the food web, rather than production being targeted toward a single apical predator, which is evidently not the case here.

Although the concentric circles in Figure 51 do not accurately represent trophic levels, the convention is adopted of showing cannibalistic species half a "trophic level" higher than otherwise, on the grounds that they share some of the characteristics of their predators. Thus Antonogadus (species no. 9) is exclusively invertebrate feeding, except for a predeliction for consuming its own kind, and is shown half-way between the first and second levels. Similarly this applies for Micromesistius (6) and Phycis (7); cannibalistic species which also consume other fish species as well as invertebrates. It may or may not be a coincidence that with species (8), these cannibalistic species (Micromesistius pontassou), Phycis blennoides and Antonogadus megalokynodon) play a key role in the diets of the apical predators, and we can speculate (Ursin, 1982) that they may be able to move upwards in the heirarchy to occupy the place of one or more of the apical predators should they be temporarily reduced in numbers.

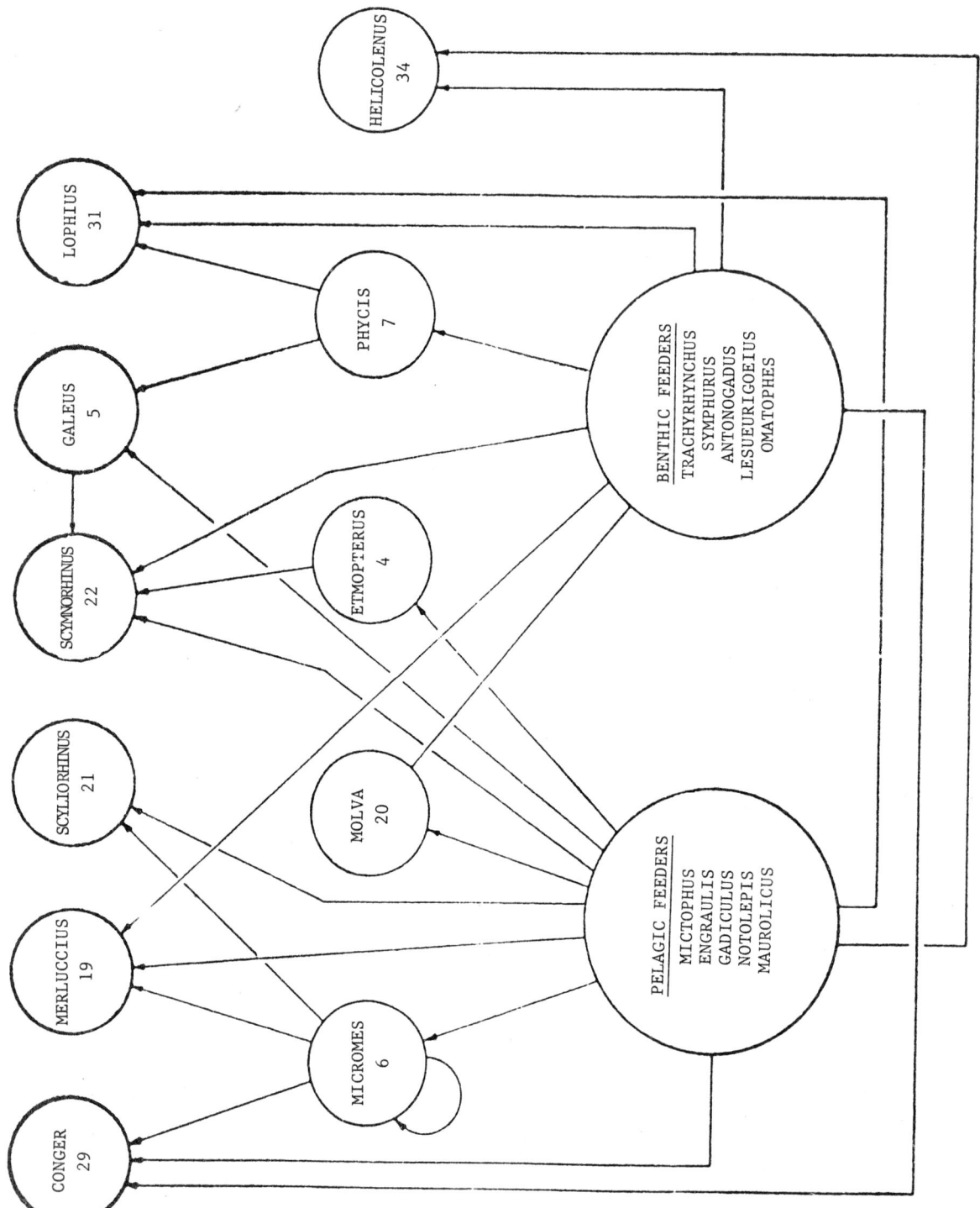

Figure 52 Compartment model for the same system shown in Figure 51, illustrating the two main groups of forage fish, classified by their predominant food type

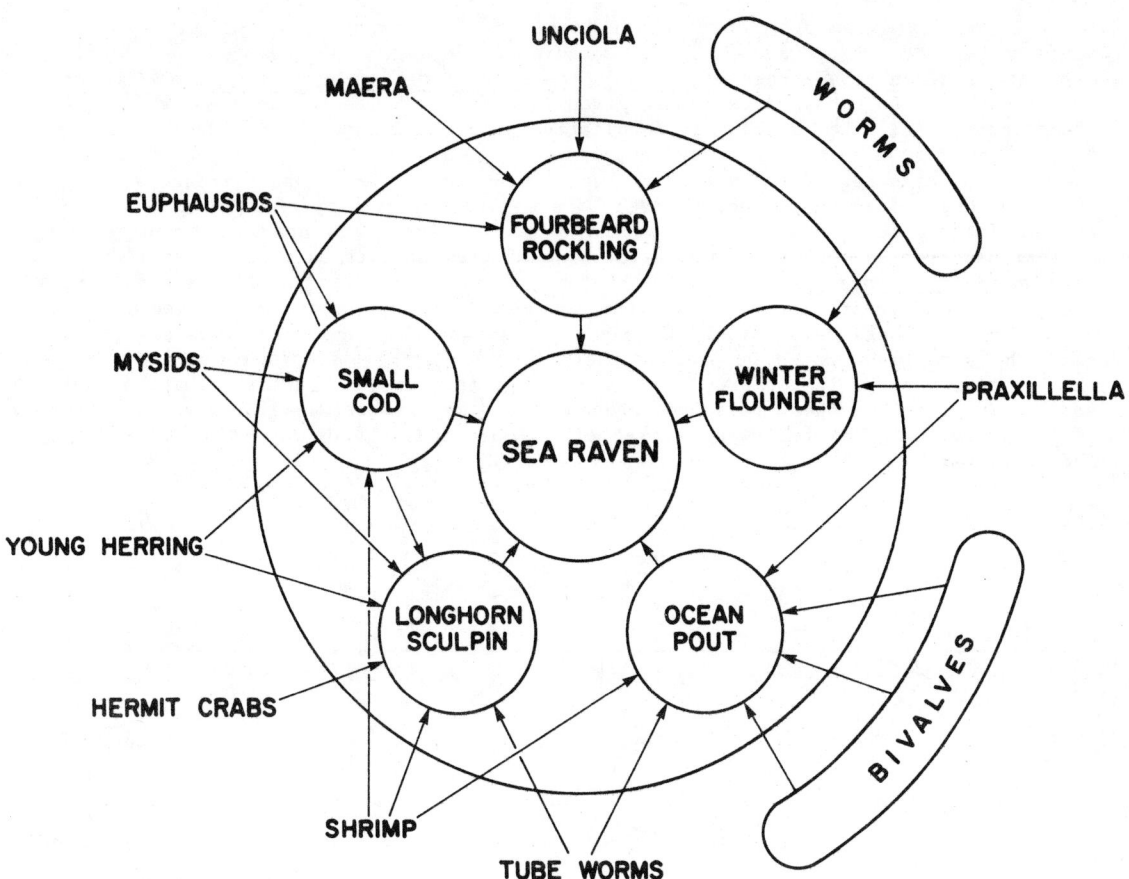

Figure 53 Showing the "Energy usage group" of the sea raven (<u>Hemitripterus</u> <u>americanus</u>): (in contrast to Figure 43, forage species are shown outside the outer circle) (From Tyler, Gabriel and Overholtz, 1982)

Several features of this food web attract our attention and may have significance:

1. Discounting the fragile distinction that a few species (nos. 6,7,8,9 and 24) are only one linkage at maximum above the level of wholly invertebrate feeders (invertebrates in fact make up a large part of their diet), there is no real distinction between arbitrary levels 2 and 3. In fact, for the most part (except <u>Molva</u>), invertebrates play an important part in the diet of all components, so that 'percent invertebrates' in stomach contents might be inversely related to 'height' in the food web, and as such used as a rough index of trophic level.

2. Also more or less as a consequence of (1), we may consider this food web as essentially very broad and shallow. Although some higher components may exist that are not shown (e.g., sharks), there appear to be few if any components wholly dependent on a single item in the diet.

3. One of the other interesting features of this food web is the relatively infrequent occurrence of cannibalism (which can be regarded either as an intraspecific population control device, or as a mechanism for population survival in limiting food conditions - see later), and the apparent absence of reciprocal predation. This last phenomenon (mentioned earlier between herring and cod) would undoubtedly be seen if larval stages were considered, but seems to be absent in the macrofaunal fish components. The explanation for this appears to be that individual size ranges of species as adults are relatively limited (compared to high-latitude commercial species) and stratification by size in this community is rather well developed with many small species, and a relatively large ratio of mouth gape to body size for the fish predators. It would be interesting

to know to what extent this is true of tropical fish communities: certainly the ratio mouth gape/body length should be some measure of the degree of "compression" of the food web into a given total range of sizes in the ecosystem. An example of the wide range of this ratio in nature has already been given in Figure 10B. (Note, however, the importance of effective gill raker spacing in facultative particle feeders, e.g., many scombrids).

4. There is a large degree of overlap between the diets of the apical predators, suggesting that the system should accommodate to a moderate intensity of fishing effort on any one predator species, i.e., there should be a relatively low probability of population explosions or catastrophes in population size of prey species as a result of reducing the abundance of any one predator in the system. This speculation is of interest in view of the major local historical fluctuations shown by catches of one of the more important predators, the hake Merluccius. We may question whether the other predators sharing in part the same prey showed population increase out of phase with hake or alternatively, whether fluctuations in the prey species were the main casual agent for the fluctuations. An investigation of these relative abundance changes with time will be very instructive in this regard. The following subset of the above food web may then have management implications:

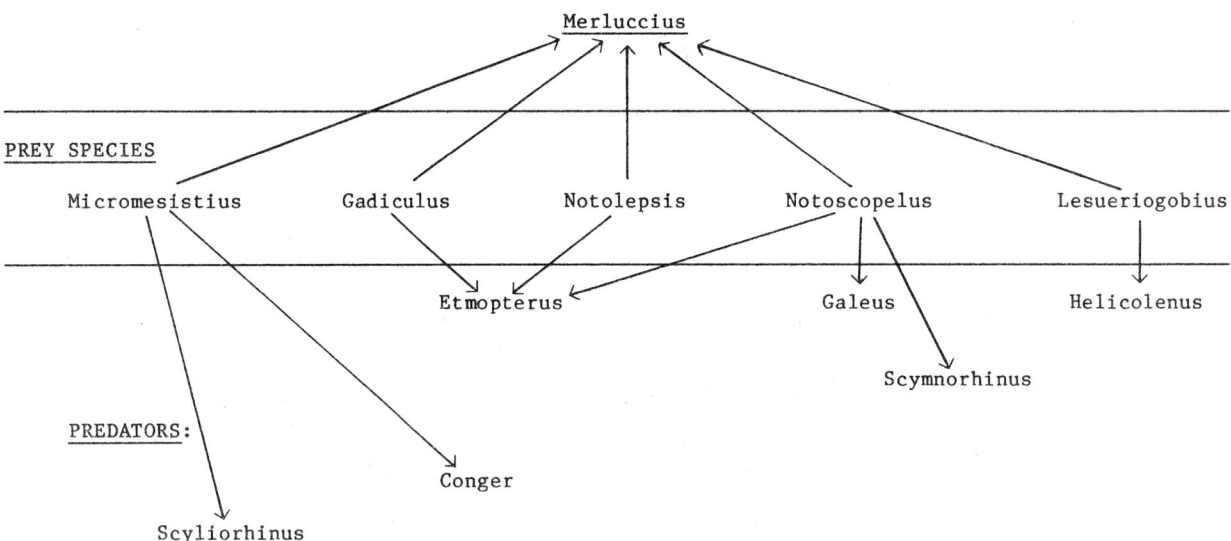

Two types of interaction may be detected between the key commercial species Merluccius and other components of the system: competition between predators, and mutual impacts of predator and prey populations. Simple measures of these two potential impacts are given in the following table, namely, the percent similarity of the diets (effectively the percentage of diet in common between two predators), and the combination of availability and "preference" (as judged by the ranked percent of a given prey species in the stomachs of hake and the other predators). Both of these criteria may be used to determine whether changes in hake population size are likely to have an impact on other directly-linked food web components.

	Ranked similarity of diet with Merluccius (%)		Ranking by importance of Merluccius as a predator in terms of the proportion of prey in stomach contents	
1	Scyliorhinus	35.4	Lesueriogobius	1
2	Etmopterus	33.0	Micromesistius	2
3	Conger	26.5	Notolepis	2
4	Galeus	10.3	Notoscopelus	6
5	Helicolenus	10.1	Gadiculus	9
6	Scymnorhinus	4.8		

Several other simple indices may be proposed, in particular, the relative consumption of a given prey by a given predator. This could be most easily measured by:

(% wt. of a given prey in the stomach) x No. (wt.) of predators.

This index should preferably also be adjusted by the rate of food consumption of the predator before comparing between species, and also ideally take into account the seasonally-adjusted percent habitat overlap of the predator and prey. In conclusion, an index of competition (I_c) between predators A + B could be postulated:

$$I_c = (\text{\% similarity of diet of A + B}) \times \left| \begin{array}{l} \text{seasonally adjusted overlap} \\ \text{in time and space of species} \\ \text{ranges for A + B} \end{array} \right|$$

and an index of potential impact (I_p) of predator C on prey D by:

$$I_p = \left| \begin{array}{l} \text{Feeding rate} \\ \text{per unit wt} \\ \text{of predator C} \end{array} \right| \times \left| \begin{array}{l} \text{Biomass of} \\ \text{predator C} \end{array} \right| \times \left| \begin{array}{l} \text{Seasonally adjusted} \\ \text{\% D in diet of C} \end{array} \right| \times \left| \begin{array}{l} \text{Seasonally adjusted} \\ \text{overlap in range of} \\ \text{C + D} \end{array} \right|$$

Comparison of the components of a food web in this fashion might detect pairs of species most likely to be affected by changes in biomass of the other. This should be helpful in preliminary risk evaluation of possible effects of intensive harvesting of one or more species on others of similar diet, or which are key predators or key dietary items for the species in question.

Ecosystems in upwelling areas

In general, Mediterranean demersal food chains such as we have just been discussing, are nutrient limited, and usually have a rather high diversity with a wide range of species sharing the biomass.

In contrast to this type of habitat we may consider the ecosystems of upwelling areas: (see Sharp and Csirke, 1983; Bas, Margalef and Rubies, 1985) for more details on these systems). Upwelling areas support some of the world's most productive fisheries. Even in low latitudes, such systems are driven by cold nutrient-rich water rising to the surface as a result of hydrodynamic forces, principally the effects of offshore winds, which are often seasonal in occurrence, as is the upwelling 'event' in most cases. Typically, the strength of upwelling varies from year to year (Figure 54), and the fish catch with it (Figure 55): making these areas very dependent on environmental factors as well as on fishing effort, for short-medium term fish production.

In contrast to other high diversity, nutrient-limited habitats such as those typical of coral reef areas in the tropics, areas of upwelling are nutrient-rich, and support a high production of phytoplankton leading to shallow food webs showing a high species dominance by relatively few resident components. Each component is characterized by a high biomass which is rather variable in time. In some respects, as noted by Cushing (1982), these systems show a similarity to the fisheries of productive high latitude areas, such as the North Sea, with certain obvious substitutions, in that the "cods" are often replaced by hakes, the mackerel by jack mackerels, and the herrings and sprats by sardine and anchovy: this is obviously a gross oversimplification, in that some groups such as the hakes also occur outside upwelling systems, while important components for example, the cephalopods and sparids are not mentioned. For most upwelling systems, this similarity is nonetheless reinforced by a seasonality of production, which also contributes to the generally low stability and diversity of such systems.

Ryther (1969) notes that the average length of food chains is shorter in oceanic upwelling areas, and this is related directly to a low diversity and an earlier successional stage than more stable systems elsewhere. Much of the production is transported away from the immediate centre of upwelling; both horizontally, to adjacent areas (Figure 56); and vertically, with depth to benthic food chains (Figure 4). Thus, it has been noted that lower trophic levels in upwelling areas are often displaced in space with respect to one another, progressively away from the centre of upwelling. Thus, off the coast of NW Africa for example, primary production is not intense close to the coast, but the greatest zooplankton production is near the edge of the shelf. For fishes and other predators higher in the food chain of seasonal upwellings (e.g., Figure 21), a degree of seasonal mobility or migration is called for, which may take them still further from the centre of production. Transportation may be passive: cold, upwelled water can be carried for at least 100 km from the point of emergence; thus forming what we have referred to as a dissipation structure (Figure 56), in which the centre of production of successively higher elements in the food webs are displaced laterally (by drift and migration) from the centre of primary production. Jones (1982) notes that the importance of this surface residence time is that primary production can be completed and assimilated into the trophic chain before the cold water sinks again: thus

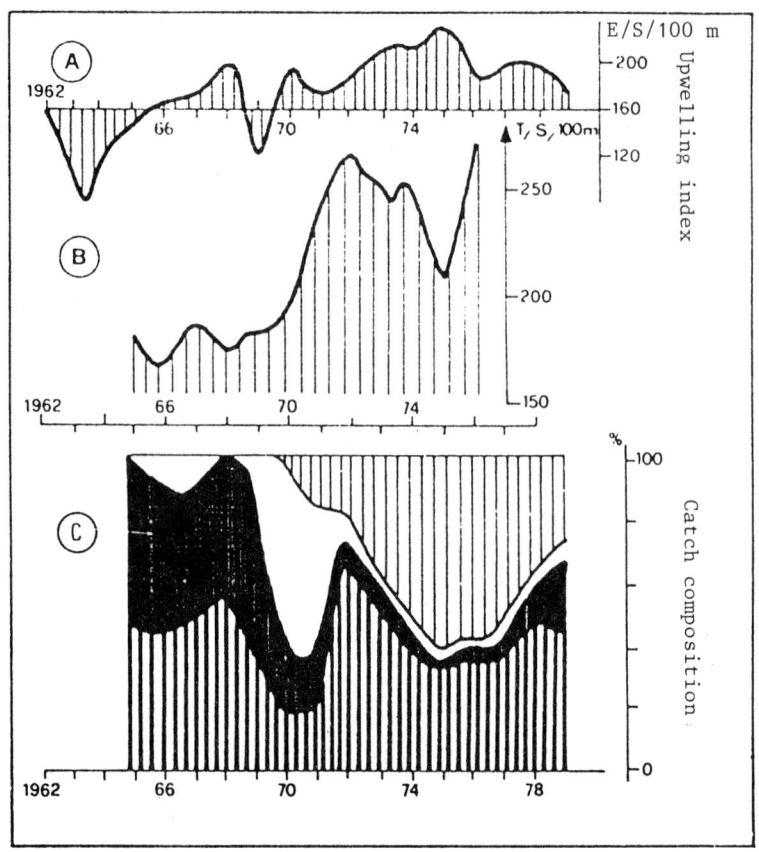

Figure 54 (A) Variations in upwelling strength in the central part of the CECAF area

(B) Variations in upwelling strength in the southern area

(C) Changes in species composition of catches

(From Gulland and Garcia, 1984)

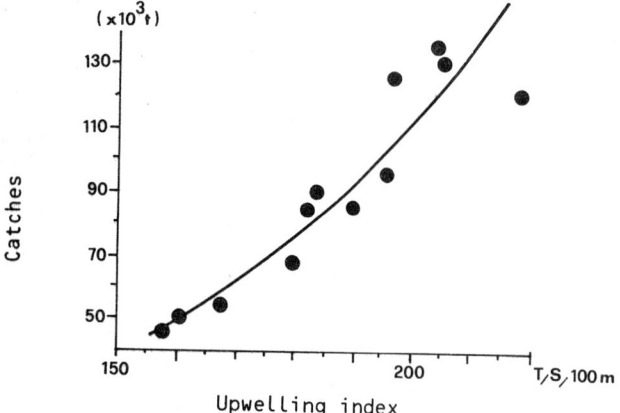

Figure 55 Relationship between annual catches and upwelling in the CECAF area (From Gulland and Garcia, 1984)

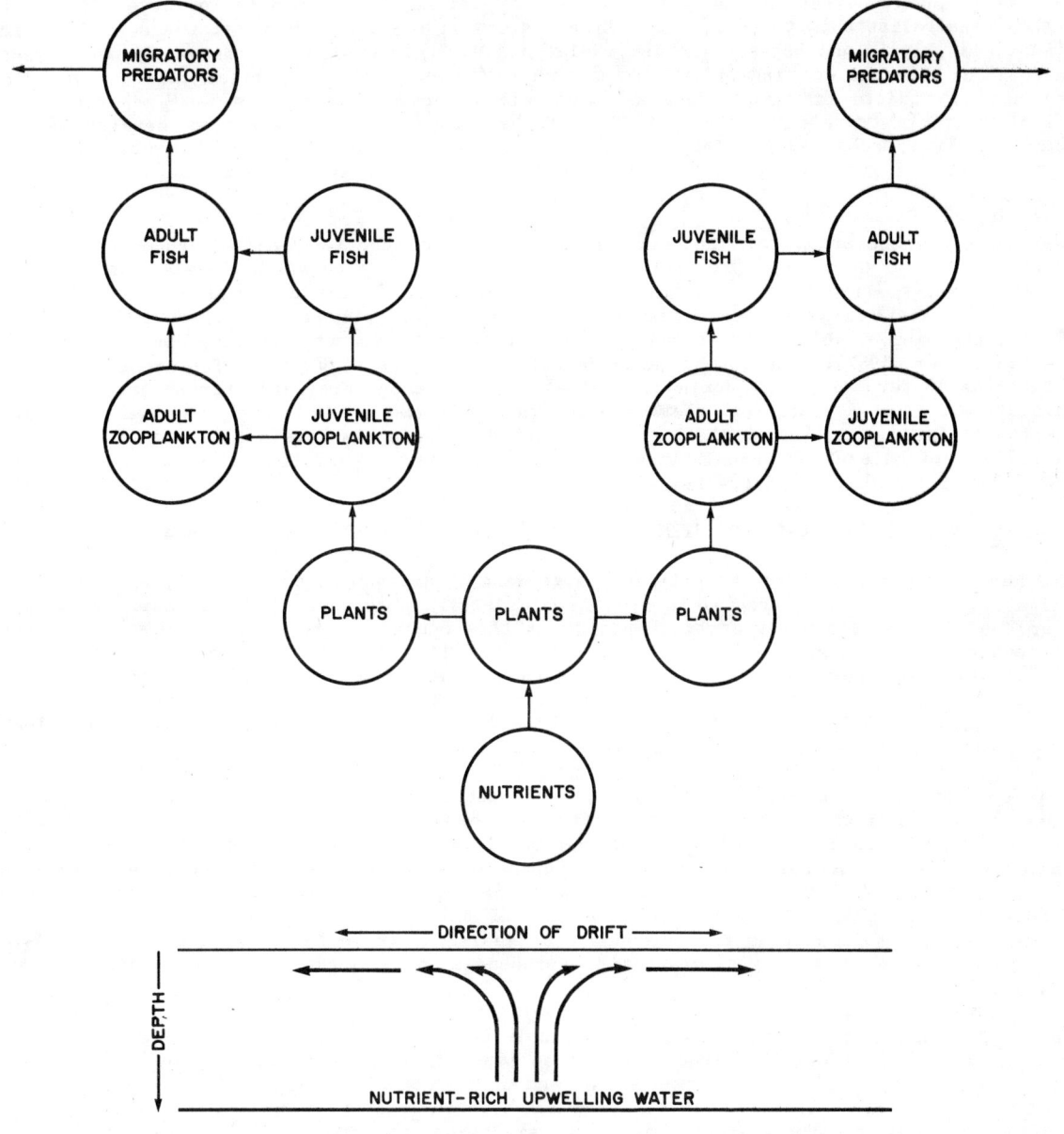

Figure 56 Dissipation structure typical of food webs of upwelling areas
(Modified after Jones, 1982)

accounting for the short food chains. By lateral advection, however, the impact of upwelling systems may extend to a considerable distance from this centre, both because of 'leakage' of nutrients horizontally and precipitation of detritus vertically. As noted, "leakage" occurs also through the seasonal migrations of many fish species, which dissipate nutrients from centres of high production. Migration is also necessitated by phenomena associated with seasonal upwellings, such as the tipping of oxyclines and thermoclines, which can result in low oxygen bottom water approaching the surface and making such waters uninhabitable for many species. Leakage occurs vertically also to the benthic systems, which may show a tendancy toward anoxic conditions due to decomposition of abundant detrital "rain" from the pelagic production system.

It follows from the generally low diversity of upwelling ecosystems that food webs are simple and that the size spectrum of organisms at any one time is likely to be somewhat discontinuous. Despite high food abundance, this could lead to problems in locating appropriate sized food particles at key stages in the life history of resident species, since size preferences of food necessarily change as a species completes its ontogeny.

One consequence of this may be the fact that several key components of low diversity environments show cannibalistic features in their life histories, e.g., squids (Amaratunga, 1983), anchoveta (Santander, 1981) and hakes (Lleonart, Salet and MacPherson, 1983): the young of the year of Merluccius (and the smaller individuals of a cohort (males), in the case of Illex), providing a "safety net" for diets of mature females, thus aiding production of a second generation under otherwise uncertain trophic conditions. Such mechanisms also would serve as density dependent population control mechanisms; "damping out" wide fluctuation in abundance in relatively simple food webs. (See Figure 57 for a rather speculative view of such a triangular component in upwelling food webs).

Csirke (1980) described how anchovy recruitment is a function of two processes: a stock dependent mechanism where egg production decreases at high population levels, and a density dependent mechanism where prerecruit survival is a function of year class size and environmental events. These year to year events include the variable degree of development of the El Niño (upwelling) phenomenon, which in turn, has a controlling influence on dominance among anchoveta and sardine (Santander, 1981): the former being associated with the high production areas close to the upwelling, the latter being more dominant in less productive oceanic water farther from the coast. In years of less intense upwelling, these species ranges become more closely coincident, and direct competition of larvae, as well as the high densities of spawning anchoveta close to the coast, must also result in high levels of cannibalism of anchoveta eggs at this time (see Csirke 1980; Sharp and Csirke, 1983 for further discussion).

13. INTERACTIONS IN A PELAGIC ECOSYSTEM: INFERENCES FROM FISH PHYSIOLOGY AND BEHAVIOUR STUDIES

We have already described the pelagic environment as a dynamic, contagiously aggregated, system which reflects diurnal, seasonal as well as historical processes. It is perhaps reasonable to consider the upper actively migratory component of this oceanic food web as system "integrators". They smooth out the discontinuities in biomass by travelling relatively quickly between abundance centres, harvesting these, and thereby reducing them in abundance to a lower "background level". Any location which is a seasonal feeding ground for migratory or nomadic predators with temporary peaks in abundance of forage species, is "smoothed" to a great degree by these visits, leaving "growing room" for all levels of the food web.

The concept of "steady-state" interactions of species in any food web could only persist if the many fluxes of material between the different components by death, emigration, growth and recruitment all balance each other out, and continue to do so in the face of human harvesting. We have already seen that reproduction or survival success is not only seasonal but annually variable in most species, and that there is little evidence for a general coordination or synchrony between spawning success of food web components. Because of this, any attempt to try and "balance" the production and consumption terms for a food web are unlikely to be fully successful, but this does not reduce the value of such a diagrammatic representation as a clue to possible dynamic interactions.

We might infer that the majority of individuals that reach large sizes in tropical systems do not come up against the kinds of food limitations imposed at higher latitudes by strong seasonal cycles of primary production, where toward the end of each winter, the available food is low, and hence growth is slowed or stopped. In the tropical habitat however, despite the continuous trophic productivity and the continuous grazing on it, quasi-steady population growth of prey with sporadic or continuous reproduction, is associated with patchy prey distribution and higher metabolic rates. Continuous recruitment to prey sizes occurs, which could ideally promote continuing and relatively rapid growth for higher predators. These may need to be highly mobile, however, to satisfy their energy needs, especially in oceanic waters, due to the relatively low densities of suitable prey sizes in many time-area strata.

In the higher latitudes the larger predators, e.g., cod, tend to disperse after the main production bloom is past, possibly simply to facilitate search and feeding outside the main centres of production. Fishes in higher latitudes tend to overwinter in near torpid states, with metabolism minimized through temperature effects. Many species select low temperatures sites for overwintering.

What is clear is the contrast between the seasonal speed-up and slow-down cycles of high latitude ecosystems, and the continuous rapid turnover of tropical systems. Whereas higher latitude predator-prey interactions have been given a lot of attention (Anderson and Ursin, 1977; Ursin, 1982), the tropical pelagic system has only recently begun to be studied in a similar fashion.

Recent studies of feeding of tropical tunas by Olson (1982) have begun to better formulate the problems and possible interactions of higher predators and their prey in the tropical pelagic-oceanic environment. Recognising that there have been several attempts to study these processes qualitatively in tuna ecology (Blackburn, 1968; Williams, 1966; Magnuson and Heitz, 1971), it is

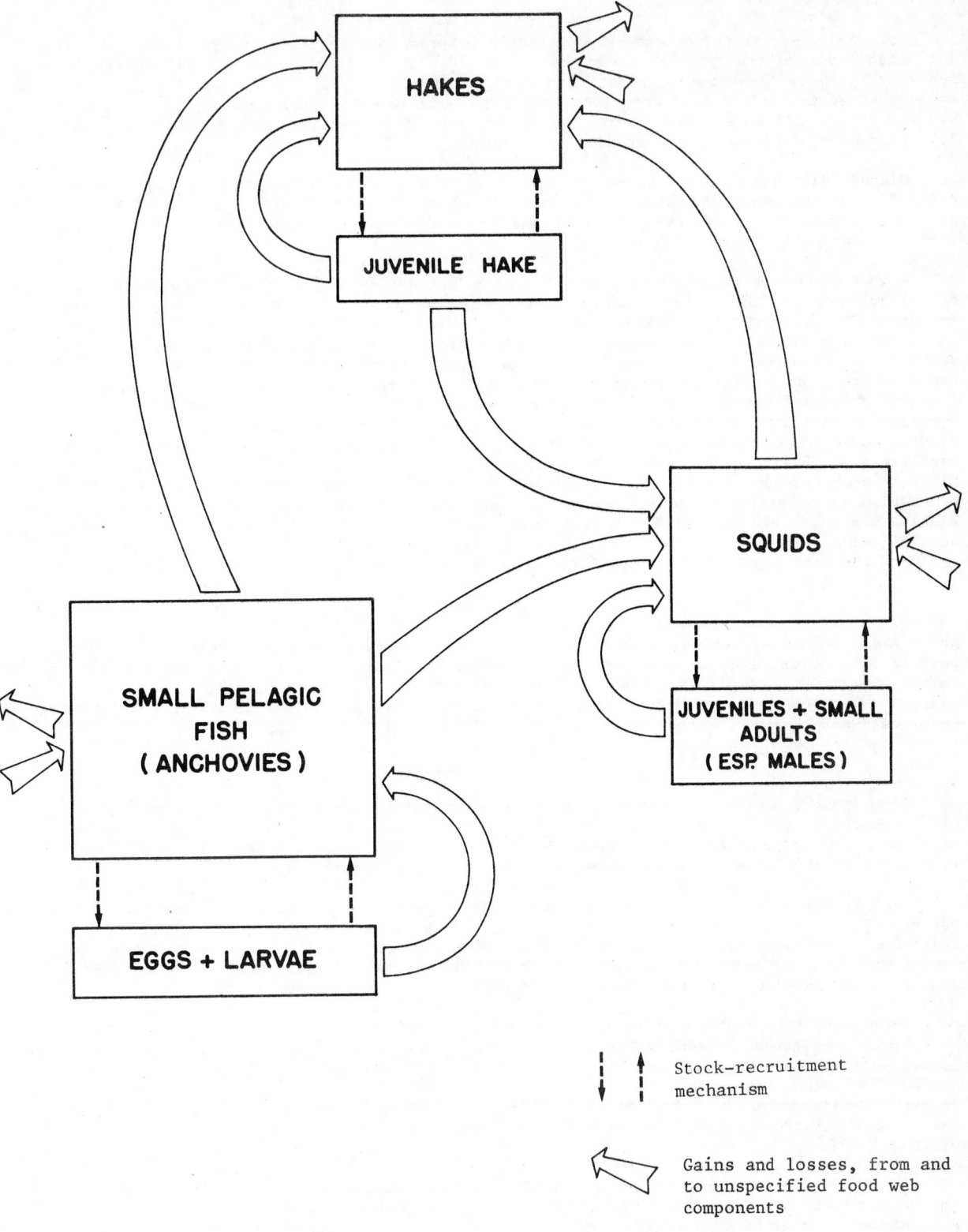

Figure 57 Generalization of interspecific and intraspecific feedback mechanisms (predation and cannibalism) for density-dependent population control in upwelling systems

satisfying to see the increased utility of this information, once quantitative information on the items consumed becomes available.

For example, yellowfin tuna are opportunistic predators which reside in the upper thermocline and mixed layer of the tropical oceans of the world. From rough approximations of their level of activity (especially swimming, growth and background metabolism), Sharp and Francis (1976) estimated that an individual yellowfin tuna (of 85 cm body length: 12.1 kg body weight), expends only 9.5 percent of its consumption energy on growth, and 19 percent on background metabolism. About 71.5 percent of food energy is expended in swimming.

Olson (1982) has followed up on this study by examining yellowfin tuna stomach contents, their calorific values, and evacuation times. Olson used an Index of Relative Importance (IRI) in integrating information on frequency (F), volume or weight (W) and number (N) of food items in the stomach, where IRI= (%N+%W)%F (Figure 58).

A methodology still commonly used today in estimating daily food intake is based on work by Bajkov (1935). In Bajkov's method the amount of food eaten during some time interval is determined by measuring the amount of food in the stomach at the beginning and end of the interval, and correcting for the amount evacuated from the stomach during the same interval. Bajkov (1935) assumed that fish feed continuously and that gastric evacuation rate is linear and constant for all types of food consumed, and suggested that the daily meal (1) be given by D = 24 A/n, where A = mean amount of food in the stomach and n is the number of hours to complete stomach evacuation. Numerous workers have since established that the rate of gastric evacuation in fishes more often fits an exponential model, and that it is significantly affected by changes in food type and food particle size (Elliot and Persson, 1978; Windele, 1978; Fänge and Grove, 1979). Eggers (1977) and Elliott and Persson (1978) show mathematically that substituting an instantaneous gastric evacuation rate "yields a correct and robust estimator" of the daily meal. If frequent stomach samples are taken over the entire day in order to detect any diel periodicity in feeding activity, and the amounts of food in the stomachs at the beginning and end of the 24-hour period are equal (Elliott and Persson, 1978; Eggers, 1977), then the daily meal D is given by:

$$D = 24 A_i R_i$$

where A is the average amount of food species i in all stomach samples, and R is the instantaneous gastric evacuation rate per hour for food species i. Since yellowfin tuna are presumed to feed only during daylight hours (Reintjes and King, 1953; Schaefer, Broadhead and Orange, 1963), stomach samples from purse seine caught fish may be used for this analysis. Assuming daylight feeding, this would only change the above equation to the following:

$$D = 12 A_i R_i$$

Olson points out that different types of food organisms may be digested and evacuated at quite different rates, and that it is desirable that gastric evacuation rates of a representative variety of natural food organisms be measured in the laboratory if the quantitative rates of food consumption are to be estimated, which is essential for studies of energetics.

Such measurements were made on captive yellowfin tuna, and the weight of stomach contents converted to calories (the calories per gram of the various food items being either determined by bomb calorimetry, or taken from the literature). A wide range of calories per stomach content was found in yellowfin at any given fork length. Estimated daily meals ranged from 0 to about 480, 2 600, 5 600 and 17 000 Kcal/day in age-classes, 1,2,3 and 4+ years, respectively.

These daily consumption figures are considered underestimates since they are based on the amount of food found in tuna stomachs after the fish had been enclosed in a purse seine for a number of hours, frozen aboard the vessel, partially or completely thawed prior to unloading, then sampled, refrozen, thawed and examined. These processes and an unknown rate of regurgitation prior to death would reduce the volume of food found in the stomach samples below that found in freshly-caught fish. Therefore, the true daily meal sizes are undoubtedly greater; but how much greater is unknown.

Although the daily meal estimates based on stomach samples are quite low, Olson used them to derive purposely conservative estimates of prey biomass consumed by yellowfin. The calories eaten per day were converted back to grams of food, and daily meal values shown separately for different age-classes in Table 11. The estimated total prey biomass eaten per fish-day was then partitioned by prey type based on the relative proportions in which they occurred in the 1970 stomach samples. These preliminary estimates suggest that a yellowfin in the 4+ age-class may eat at least 730g of frigate tunas (7.3 individuals of 20 cm average size), 440 g of nomeids (146 individuals of 3 g average weight) and about 233 g of other prey per day. Yellowfin in age-class 3 appear to eat an average of at least 235 g of frigate tunas (2.4 individuals), 100 g of nomeids (33 individuals) and

Table 11

Estimates of prey biomass eaten by the CYRA (Commission Yellowfin Regulation Area)
yellowfin tuna population per day and per year in 1970 derived by multiplying daily meal
(prey biomass consumed per fish-day) by the number of individuals in four age-classes estimated by
cohort analysis

	Age group				
	1	2	3	4	Total
Prey biomass eaten fish^{-1} day^{-1} (g)	88	730	461	1 403	
Number of individuals in CYRA (Sharp and Francis 1976)	31 731 100	14 257 730	6 406 050	3 661 630	56 056 510
MEAN PREY BIOMASS EATEN BY YELLOWFIN POPULATION/DAY (in t):					
Scombridae		1 161	1 505	2 686	5 352
Gonostomatidae (Vinciguerria lucetia)	1 221	426	122	180	1 949
Nomeids (Cubiceps sp.)		352	649	1 608	2 599
Cephalopoda (squids, octopus, argonauts)	1 151	372	197	88	1 808
Exocoetidae (flying fishes)	87	346	239	179	851
Pleuroncodes planipes (red crabs)	264	155	113	102	634
Unidentified fishes		75	75	88	238
Balistidae (trigger fishes)			21	69	90
Miscellaneous	69	30	41	138	278
Total prey biomass eaten/day (in t)	2 792	2 917	2 952	5 138	13 799
Prey biomass eaten/day (in '000 t)	1 019	1 065	1 077	1 875	5 036

about 125 g of other prey per day. Fish in age-class 2 eat at least 80 g of frigate tunas per day on the average (Figure 58).

An important caveat needs to be added here. This data set and analysis are based on mean stomach contents assumed to be maintained over a 24 hour feeding period. We do not know what the relative proportion of feeding is between night and day for this species, but feeding rate is likely be less at night than that determined from these daytime collections. In this case, the proportional daily consumption rates would be less than those presented. However, if maximum rations at each size grouping are used, (all stomachs as full as the fullest in each size class), Olson (personal communication) estimates that the four size groups would consume 16.0%; 15.7%; 19.0%; and 16.7%; per day using the 12 hour feeding model and 32.0%; 31.3%; 38.0%; and 35.5% per day using a 24 hour feeding model. The 12-hour rations model corresponds exactly to the results of feeding experiments by Magnuson (1969) in which skipjack tuna were offered food *ad libitum* over 12-hour periods. Olson remarks:

"One can begin to get a feel for the tremendous numbers of frigate tunas (*Auxis* sp.) that are consumed by the entire yellowfin population". Some rough estimates were made of food consumed by the yellowfin population of the Commission Yellowfin Regulatory Area (CYRA) in 1970 (Table 11). The per-individual biomass consumption estimates were multiplied by the number of individual yellowfin in each age-class in the CYRA as calculated by cohort analysis (Sharp and Francis, 1976). These estimates indicate that at least 13 799 metric tons of food were eaten per day by the population: 5 352 metric tons of which were frigate tunas (= about 53.5 million individuals). A total

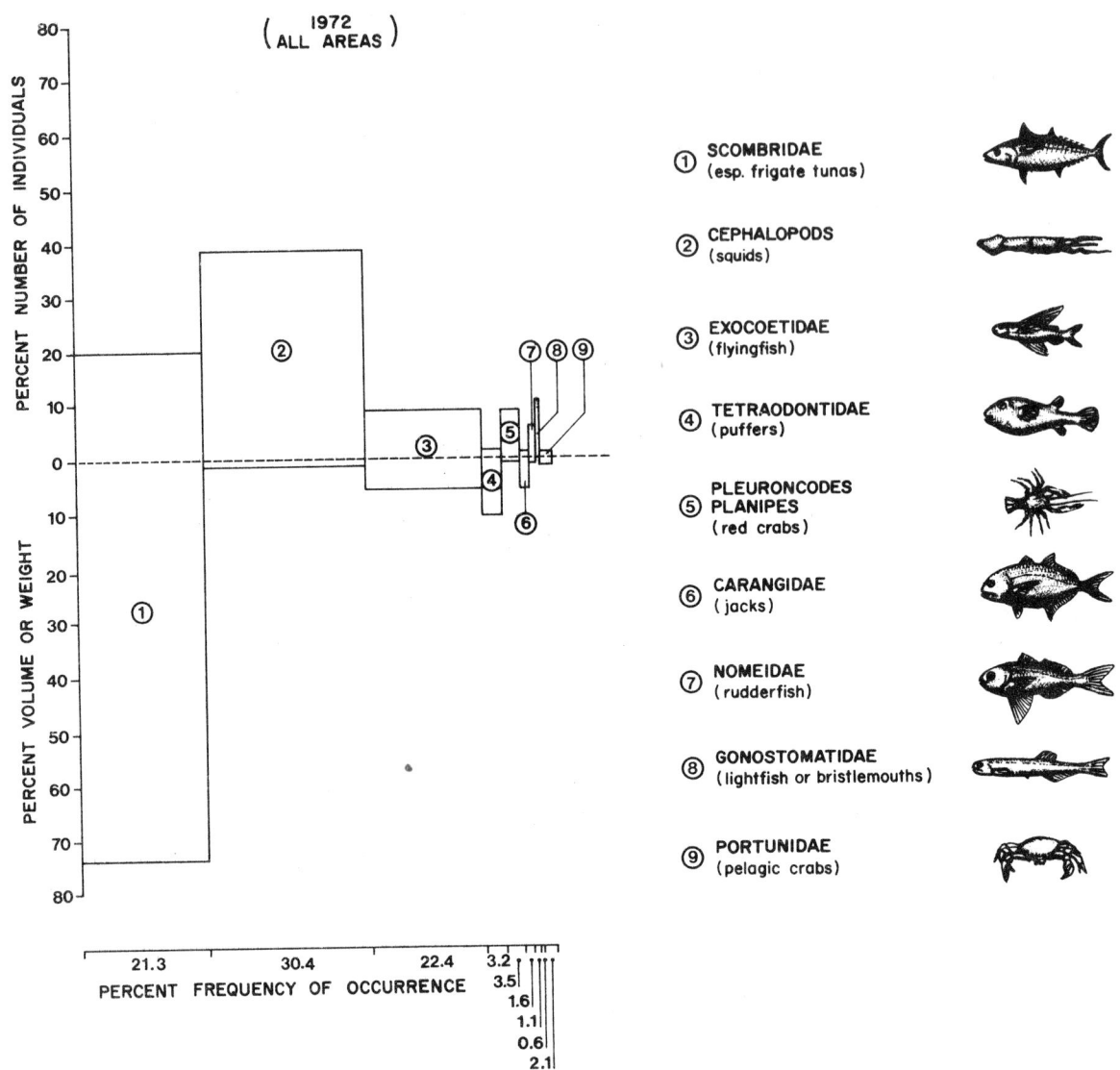

Figure 58 Showing relative proportions of different food items in the stomach contents of yellowfin tuna in the eastern Pacific (From Olson, 1982)

of at least 5 million metric tons of food were consumed during the year, about 2 million tons of which were frigate tunas. Again, these are undoubtedly underestimates of the tremendous numbers of Auxis consumed by yellowfin tuna annually.

The significant new insight into tuna energetics gained through detailed analyses of this type is that:

(a) yellowfin have an overall ecological efficiency (biomass consumed to biomass growth) of about 0.15 percent, due to their extremely expensive method of locomotion (Sharp, 1983);

(b) the amount of food consumed by tuna (and hence the production, and even the standing stock of prey) must be very high. This also has serious implications for the real levels of production in offshore areas, which may have been significantly underestimated.

Of course, all of this needs to be kept in perspective. In a related approach Sharp and Dotson (1977) used fat content measurements and hydrodynamic experiments to determine that the flux in fat content in albacore could be accounted for by the energy expended in "directional migration" alone, while any day-to-day activities would be accounted for within the daily diet and energy conversion.

The question of resident versus migrant components has been bothersome in studies on many widely distributed resources. Fat content and growth rates can be useful in identifying new migrants, i.e., what proportion of available fish at any one time might be entering grazing/fishing areas. Values for fat content ranging from 2 to about 18 percent of total body weight have been measured in albacore (Dotson, 1978; Sharp and Dotson 1977), indicating the dynamic nature of fat reserves, and perhaps also the mosaic nature of populations. Coastal fish of 63 cm average length had lost an average of 404 g of fat on entry to the fishery compared to fish of the same size sampled about 1 000 nautical miles offshore prior to the commencement of the coastal fishery. Two weeks after their initial arrival on the coast, fish of similar size were sampled again in the coastal fishery and their fat content was back in the range of the offshore material. These observations obviously fit well with the role of highly mobile predators as large system integrators.

The above account can be compared with a similar and equally interesting series of studies on larval fish swimming energetics. Vlymen (1974, 1977) showed that Engraulis mordax (California anchovy) larvae have a relatively high proportional expenditure of their total respiration due to swimming, i.e., 24.6 percent under average activity level conditions for a 1.4 cm larvae. Smaller anchovy larvae are less efficient due to their sporadic swimming behaviour, but efficiency increases as they grow larger and settle into their pump-and-glide mode of swimming.

Vlymen's (1977) analysis of food microdistributions, growth and behaviour for anchovy larvae illustrates immediately the relevance of scales of turbulence and other oceanographic discontinuities to survival of larval fishes. (These concepts are further examined in IOC Workshop Report No. 28). There are discontinuities and contagion of particles and organisms in the marine environment (Sheldon and Parsons, 1967; Sheldon, Prakash and Sutcliffe, 1972; Platt and Denman, 1975, 1977; Owen, 1980), which have profound effects on local processes, and particularly on local abundance of the primary through tertiary predators some time after the initial bloom in primary production has occurred (Sharp, 1981b). In general, the scale of events in space and in time in food webs in the ocean are correlated (Figure 59; Carlenton, 1985). The relative energetic efficiency of the more mobile, secondary and tertiary predators, e.g. the tropical tunas, which search for one bloom of prey species after another, is far lower than for larval fishes, or even for adult neritic fishes. This is shown for example, by Sharp (1983), Sharp and Dotson (1977), and Sharp and Francis (1976).

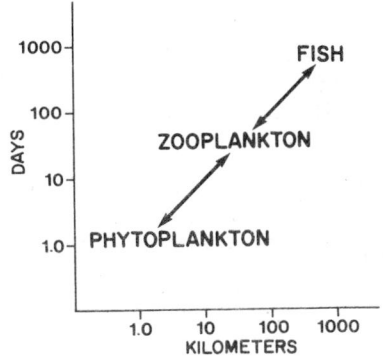

Figure 59 Scale relations in the pelagic environment among phytoplankton (P), zooplankton (Z), and most adult fishes (F) (Redrawn from Carlenton, 1985)

In reviewing the topic of energetic interactions with the intention of adding ammunition to the stockpile of fishery tools, it became clear that there are several semi-mythological concepts in fishery science which need scrutiny, particularly in relation to predator-prey effects.

One of the first of these is the assumption that harvesting the adults of species A will a priori make room for more younger, faster-growing, more productive members of the same species. This tenet will not always prove to be correct, and stems from the concept of "carrying capacity" inherent in most ecological and fishery population models. This assumes a finite biomass of individuals of a given species can be supported by the environment. This situation is only approximated when either a stable age distribution, hence fixed requirements exist, or when the population is comprised of identical individuals, e.g., single cells in a stable, nutrient-rich environment. These assumptions may be incorrect for fish populations in nature due to the behavioural and energetic discrepancies between the various species and size/age groups of most fish. Adults are rarely in direct competition with juvenile or earlier stages, and may occupy different loci in the food web. Except for scombrids which do not exhibit such features as a gas bladder or fat deposition, e.g. Katsuwanus and Euthynnus, and hence have higher energetic requirements with increased size, fish growing to large sizes, i.e., over 10 kg, may actually be more ecologically efficient than smaller individuals. The more usual case however, is that any vacuum resulting from harvesting species A will be filled by another species, let's say species B, with higher reproductive capacity, faster growth, and a generally more efficient energetic turnover rate, i.e., more ecologically efficient.

These comments of course beg the issue as to just what the "carrying capacity" limiting the size of any species might be. It might only be reached in the adult stages during periods of low or zero fishing, and extremely high abundance. The more usual "population limitations" may however be fixed by the available "survival habitat" for larval fish. In most years the carrying capacity (K) for the species may simply have no sensible adult limit, or may severely constrain only the larval or early juvenile stages, such that fishing the adults has little substantial impact on changing population productivity.

The assumption often employed in simple single species fishery models, that fish species live in a biological vacuum with few or no effects on other species of relevance to management, is less easy to make nowadays. Fluctuations of stock size noted for predatory, migratory species such as bonitos (Sarda species), mackerels (Scomber species), barracudas (Sphyraena species) certainly mean that there may be major fluctuations in natural mortality of their prey species, at all size and life history stages.

This brings us back to the yellowfin and Auxis puzzle described by Olson (1982) and Sharp (1983a).

As an example of potential predator feedback effects in oceanic systems, the studies of Olson (1982) show that for the year 1970, the exploited yellowfin tuna in the eastern Pacific Ocean consumed at a minimum 13 800 t of forage per day, comprising about 2 000 t each Vinciguerria sp. and cephalopod species, over 2 500 t of nomeids and over 5 300 t per day of another tuna, namely Auxis species. The Auxis alone represents about 2 million tons per year of a major oceanic predator consumed by the exploited yellowfin population. (this is equivalent to the entire tuna harvest by man every year). The Auxis biomass is likely to be of the order of 5 million tons or more in the eastern Pacific, and harvesting yellowfin leaves an additional million tons in the ecosystem to be eaten, or to eat other species. Auxis are certainly among the more voracious and ecologically inefficient of the scombrids, consuming well over 20 percent of their biomass per day. so that we can perhaps expect as much as 70 million metric tons of scombrid food to be consumed by the extra Auxis surviving to forage, due to the yellowfin harvest (this is largely equivalent to man's total annual harvest from the seas: FAO, 1984). Since Auxis are suspected to be cannibalistic and/or eat yellowfin tuna larvae or juveniles, we might reflect on the impact of feed-back loops between Auxis and T. albacares populations due to man's removals of adult yellowfin tunas: paradoxically, harvesting T. albacares should result in a reduction of its recruitment because of better survival of Auxis!

Auxis are indeed capable of preying on larval yellowfin, but their feeding habits in the eastern Pacific have not yet been studied. Yokota et al. (1961) examined stomachs of 30 Auxis thazard caught in Japanese waters in 1959 and found that their diet included young Auxis and skipjack tuna. Squid are also described by Arnold (1979) as "voracious predators feeding when young on macroplankton and subsequently on small fish". Hurley (1976) theorises that larval squid are probably important predators of larval fishes. We should begin looking at the feeding habits of these important members of the pelagic food chain leading to the tunas in order to identify the possible predator-prey relationships that are so important to natural population fluctuations (Olson, 1982); (see also Caddy, 1983 for details of cephalopod interactions).

Probably more important is the fact that Auxis efficiency as a converter of food energy into biomass, is even lower than for the yellowfin tuna discussed in an earlier section. This suggests that besides lowering the mean size of predators in the eastern Pacific by intensively harvesting the larger yellowfin tuna, man may also be decreasing the overall production of desirable species like yellowfin tuna by shifting the ecosystem to a more "dissipative" mode, and therefore to less ecologically efficient species such as Auxis. Whether this fundamental conundrum can be solved in practise, and Auxis, (or Vinciguerria, nomeids or cephalopods for that matter), can be harvested, is another question. It is obvious that major resource biomasses exist, and may even have increased in size since harvesting the high value tunas began, but it is also evident that the energy used in harvesting these smaller species would be far higher, and their potential for marketing less promising than for tunas.

Perhaps one of the most fundamental changes in recent years in our perception of the application of food webs to commercial fisheries, is the realization that although the biological productivity of marine environments is higher than previously estimated due to the role of smaller photosynthetic cells (cyanobacteria), whose abundance has not previously been measured by oceanographers, a significant part of the energy of the food web passes through components not usable by fisheries, e.g., through "jelly" organisms such as Salps and Medusae. These gelatinous predators are not adequately represented in most sampling gears, and are quite capable of absorbing, without dramatic changes, the food energy no longer used by seriously depleted commercial stocks of smaller pelagic fish.

Together with the importance of recycling of energy by e.g. detritivores, this reinforces the conclusion that food webs centred on commercial species in nature as opposed to lab experiments, must be regarded as open ended or 'leaky', with the real possibility that even after the main linkages in the food web have been elucidated, some as yet undefined trophic linkages could be significant. It is also clear that new pathways in these food webs can become important as a result of interventions in the system.

The question of feed-back loops in marine ecosystems can only be addressed properly by sending readers to recent relevant literature. To summarize is to lose the subtle and remarkable nature of these processes. Perhaps the best point to start the survey is with the article by Ursin (1982), in which he examines the relative stability of marine ecosystems. He discovers that "triangular" food web subcomponents actually can be great stabilizing influences. These occur at all levels from unicellular or colonial predators, up through fishes, and illustrate that at each level in food webs, abundance of individual components is locally constrained. Some examples of these (from Ursin) are given in Figure 52. In the fisheries context, it is important to note that as the dominant predator is fished down, there is normally scope for the subdominant fish predator to increase in size and (partially) replace it. This is one of the reasons why removing top predators has not resulted in the startling increases in yield of their forage fish predicted by simple food web models.

Fundamental Structure

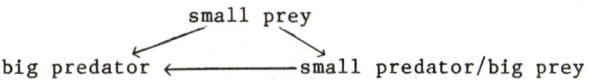

Examples:

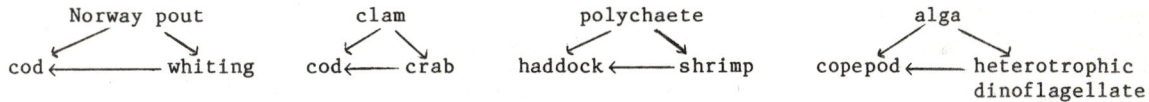

Figure 60 Triangular food-web components in marine ecology (After Ursin, 1982)

What is also evident from these triangular food webs components is the dominant impact of size in modelling food webs. (See Borgmann, 1983 and Figure 46).

Jones (1982; 1982a) considered the situation where individual organisms may find themselves located outside their "best" habitat and are subject to higher predation than the rest of the population: this provides some degree of feedback to compensate for short-term changes in relative abundance.

In functional terms, omnivory (here used in a more general sense, of the consumption of food from two or more trophic levels), should help to stabilize marine food webs. Here we are back considering the impact of predator biomass on their prey as overriding food limitations on predator biomass. Predator control might apply when predator biomass is large relative to prey biomass, but also if the predator is not totally dependent on one or two food organisms. Alternatively, the controlling impact of prey biomass is likely to be greatest when the predator is an obligate feeder on the species in question. In their simplest form, these reciprocal effects are adequately modelled by developments of the Lotka-Volterra predator-prey cycles; e.g., Shirakihara and Tanaka, 1981).

The problem of finding food efficiently when production shows seasonal peaks is solved in a number of ways: **migratory feeding** (moving from one peak prey density to another, which is temporarily and spatially segregated); **opportunistic feeding** (moving from one target prey species to another, depending on abundance), and **omnivory** (feeding on more than one trophic level and/or size range). Other alternatives are to feed intermittently (e.g., intertidal organisms), or to adapt to feeding on large meal sizes at low densities (e.g., deep water and abyssal species, which often can consume organisms of close to their own size), or to become commensal symbiotic or parasitic (e.g., remoras being carried by sharks from one meal to the next).

Another 'grand scale' ecosystem model other than that of Anderson and Ursin (1977) which is of general interest, is that being developed by Taivo Laevastu and associates (Laevastu and Favorite, 1980; Laevastu and Larkins, 1981), in which both environmental factors, fishing and predator-prey bookkeeping are attempted, and where population rates (of growth, mortality, etc.), are made functions of biomass change as opposed to changes in numbers as in most population models (i.e., ponderal rates are defined). The magnitude of competition between man's harvests and fish consumption by other marine mammals, is one of several stimulating results of this approach, showing how conflicts of interest arise as complex feed-back loops in marine systems. For example, efforts to rehabilitate marine mammal populations can only decrease the harvestable fish portion of the high latitude ecosystems for human consumption, implying a societal choice. The physical principles are simple enough, but defining the desireable 'point of balance' from a socio-political point of view is much less so, and the various objectives need to be considered in toto, not just as separate, unrelated "problems", each to be resolved by special interests.

Recognition of the complexity of interractions in the sea can only help in ordering our thinking processes about what to expect from the sea in the way of a harvest from a system with continuously changing state variables. Ursin's "relative stability" does not imply steady-state thinking is always more appropriate; it implies only that energy is not wasted.

14. CONSIDERATIONS OF ECOLOGICAL EFFICIENCY AND LIFE HISTORY STRATEGIES

Quite simply stated, the total productivity of an ecosystem is the sum of the productivities of the components. However, this statement in no obvious way simplifies the procedure of integrating trophic levels and their products in order to arrive at a single figure for the productivity of a given system. The problem lies in the differences in the efficiencies of the various inter- and intra-trophic level components, their distributions in time and space, the typical number of interactions that occur within the food web, and the life history pathways followed. The variable efficiencies of various trophic groups require first, a series of comparisons and contrasts of distributions, behaviours, and life histories.

To deal realistically with ecosystems one must recognize the inherent variety and plasticity of biological responses to physical-chemical perturbations. It seems unlikely, especially in the tropics, that the classical terrestrial concept of successional change can be uncritically applied, but each change in the system will be seized by one or more species as a one-time opportunity which will successively affect all future states. Fortunately this aspect of ecology is blessed with abundant recent reviews, and we therefore defer comments by recommending that those interested in:

(a) development of species diversity, consult Huston (1979);

(b) development of food webs, consult Odum (1968), Pomeroy (1974) and Platt, Mann and Ulanowicz (1981); and

(c) biological adaptability and ecosystem stability, consult Conrad (1972, 1976) and Johnson (1981).

Ecological efficiency

Ecological efficiency is usually expressed in terms of the change of biomass of a species compared to the weight of biomass consumed. This can be calculated in terms of calories, dry weights, or wet weights.

The efficiency of any ecosystem component depends upon several factors:

(a) abundance and concentration of each component's food sources in relation to its own abundance and concentration;

(b) vulnerability of these food sources;

(c) work done to encounter (search for, or blunder into) and capture these food sources;

(d) proportions of food sources which are assimilated into useful energy, growth, and/or vital nutrients;

(e) the proportion of the assimilated food source available to each component for somatic or reproductive growth;

(f) the complexity of somatic organization of the organism itself.

Any individual component's energetic requirements, i.e., its utilization of energy or foodstuffs, can be described in this fashion:

$$I = M + G + E$$

<u>where I</u>   is energy input in the form of light, proteins, carbohydrates, fats, which can often be monitored in the transfer of trace elements, nitrogen, carbon, oxygen, etc., depending upon the organisms or systems being studied;

<u>where M</u>   is metabolic energy which may usefully be divided into numerous kinds of metabolic work resulting from internal biochemical conversion and leading ultimately to external work, displacement heat and simple turbulence remaining for a short period after an organism has expended work against gravity, or has moved. (Note: the same symbol M, is used by field biologists for Natural Mortality rate: illustrating yet again the lack of standardization of concepts between field and laboratory biologists);

<u>where G</u>   is growth in terms of biomass, either of somatic or reproductive products, and is the only variable which can be either positive or negative; and

<u>where E</u>   is excretion, secretion or sloughing off of biogenic materials, (e.g., crustacean exoskeletons).

The energy requirements for either optimal or maximal energy flow through an organism depend to a great degree on the system characteristics, particularly for fishes (poikilotherms). Temperature and available oxygen (also environmental carbon dioxide or hydrogen sulphide levels depending upon the organism), determine the respiratory limits of the majority of living aquatic forms. Although it is not our intention to discuss here questions relating to the field of physiological ecology, it is worth noting that the oxygen uptake or respiration of an ecosystem is a direct measure of its rate of biological production (McNeill and Lawton, 1970). In response to the variations encountered in these variables, diverse adaptations and specializations have evolved to assist or buffer against stress, such as elaborate respiratory surfaces, respiratory pigments, and biochemical/genetic and circulatory specializations. The abilities of organisms to persist in uncertain and stressful environments will have been determined by adaptations to environmental gradients, and to the patchy distributions of food organisms. The storage of excess energy in the form of carbohydrates (e.g., glycogen), plus lipids and proteins for use in metabolic work, increases the organisms' resilience to varying food availability, as well as fulfilling various stasis requirements; e.g., buoyancy increases with lipid content.

The importance of motility or mobility at different life history stages of the various food web components is rarely given adequate consideration. In fact, the relevance of time and distance scales of biophysical influences has only recently been given appropriate hierarchic place in the ecosystem framework (Fasham, 1980; Owen, 1980; Sharp, 1980a; 1980b). Simpson <u>et al.</u>, 1979; Iles and Sinclair, 1982; Bakun and Parrish, 1980; Parrish, Nelson and Bakun, 1981). Perhaps the most informative place to start an evaluation of ecological efficiencies, is with an examination of energy expenditures on mobility with respect to size of organism. As noted already, this often though not always, relates to the organisms' trophic position.

From Rheinheimer's (1980) review of marine micro-organisms we find that many marine bacteria are flagellates. The typical marine bacterium has a temperature optimum within the range $18^\circ-22^\circ C$, although a full range of optima exist including facultative psychophils with temperature optima as low as $0^\circ C$. Bacteria are usually aerobic but are facultatively anaerobic in most cases. There are few organic substrata from which they cannot derive nutrients, and inorganic compounds can also be utilized as energy sources by some. One important source of nutrients is the leakage or excretion of nutrients needed for bacterial growth from primary producers (phytoplankton) as well as from any other organism, living or dead. The activities of individual bacteria are restricted to small scales, but these are not major limiting factors in their distribution. Obviously, small particles are primarily dispersed by turbulent mixing at all scales, up to ocean-scale currents. This provides opportunities for bacteria to encounter "local" food sources, as well as making bacteria accessible to organisms that feed on them.

Bacteria are supreme opportunists, with among the highest potential conversion efficiencies possible: of the order of 50 to 60 percent. Marine bacterias are found in highest densities in coastal zones, and in sediments, where many benthic organisms directly depend upon them for nutrition. Copepods also other small crustacea (eg. lobster larvae) in the zooplankton feed in part on bacteria (Rheinheimer, 1980). Cycles of abundance and scarcity due to grazing are readily observed in bacterial, as in predator populations. Benthic cyanophytes are grazed by rotatoria, nematodes, crustaceans and others micro- and meiofauna, and planktonic cyanophytes are eaten by plankton and fish as supplements, but are of decreasing importance as the predators increase in size.

Many phytoplankton organisms are not motile, and therefore their survival is a strict function of the local conditions in which they find themselves, i.e., light, nitrogen and other essential nutrients, local turbulence and system level transport must be in acceptable ranges. In contrast, there are motile phytoplankton, (flagellates), and even some bacteria, which are theoretically capable of resisting turbulent dissipation or disruption at low energy levels. All of these species depend to a great extent upon system transport physics for their distribution, physiological requirements, and for encounters with their food in order to remain productive. In this sense they are very "efficient" as individuals, sampling a vast, uncertain volume, and do not need to dissipate large proportions of their metabolic energy in swimming, as do many species at higher trophic levels.

Larval fish and planktonic stages of other organisms may graze heavily on unicellular and colonial phytoplankters. Their conversion or production rates vary enormously, depending on species mobility (Theilacker and Dorsey, 1980), and on such factors as the distribution or contagion of appropriate food particles (Vlymen, 1970; Beyer, 1981; Beyer and Laurence, 1981). The quantification of productivity in fish populations is not dealt with easily in the larval through adult stages, since once the post-larvae and juvenile stages become mobile, they become very difficult to sample. Many larval fishes are also difficult to sample due to their cryptic or very dispersed distributions.

An interesting illustration of the importance of detrital recycling in the planktonic environment relates to the small crustacea making up such an important component of these systems. Lasker (1966) notes for the euphausid shrimp _Euphausia pacifica_, that moulting occurs every three-eight days depending on temperature, irrespective of the amount of food eaten. The contribution to detritus made by the molted exoskeleton is 10% of the animals (dry) body weight: i.e., equal to the average biomass of the euphausid population every fifty days: a figure of 1.5 g/m^2 has been mentioned. Together with cast exoskeletons from other planktonic crustaceans, this component must form a significant share of the energy budget of detritivores, marine food chains, in addition to faecal material and food particles lost in ingestion.

There are many differences in modes and relative abilities of swimming in planktonic organisms, from passive drift, to near independence of local physical transport. For example, Vlymen (1970) calculated the proportional energy expenditures of _Labidocera trispinosa_, a calanoid copepod, due to vertical migratory behaviour, and to those short-burst accelerations characteristic of avoidance behaviour. The relative value of total expenditure of energy due to activity as a proportion of total respiration, is of the order of 0.1 to 0.3 percent. These incredible efficiencies are directly related to the high accelerations that copepods exhibit.

In a similar series of studies on larval fish swimming energetics, Vlymen (1974, 1977) showed that _Engraulis mordax_ (California anchovy) larvae show a relatively higher proportional expenditure of their total respiration budget due to swimming; 24.6 percent under average activity level conditions for a 1.4 cm larvae. Weihs (1980) also showed that the reason that just-hatched Engraulid larvae swim in bursts of activity for the first few days, is that this is a more efficient mode for fish under 5 mm total length. Above this size, the pump-and-glide mode becomes more efficient than continuous swimming. However, the carangiform species, i.e., scombroids, are more efficient if they sustain continuous swimming motion than are the anguilliform fishes (eels), hence the different modes of swimming for engraulids and mackerels.

Vlymen's (1977) analysis of food microdistribution, growth and larval behaviour for anchovy larvae, brings one immediately to the relevance of scale, and disruptive turbulence phenomena on survival of larval fishes. As we have said, there are widespread discontinuities in distribution of particles and organisms in the marine environment (e.g., IOC Workshop Report No. 28. Sheldon and Parsons, 1967; Sheldon, Prakash and Sutcliffe, 1972; Platt and Denman, 1975, 1977; Owen, 1980). These have profound effects on local processes, and particularly on local abundance in the primary through tertiary predators some time after the initial bloom in primary production occurs (Sharp, 1980b). The relative energetic efficiency of the more mobile secondary and tertiary predators which search from one primary bloom to the next, is far less than for larval fishes, or even for adult neritic fishes.

Caloric conversion efficiencies will depend on the energy dissipated due to swimming, on the caloric or nutritional value of the food, and on the food distribution. Laboratory experiments on two _Engraulis_ species show the adult efficiencies of caloric conversion to be about 12 percent (Hunter and Leong, 1981). These authors also discuss the proportion of this food which is partitioned among growth and reproduction by age classes of female northern anchovies. A condensed table is given below:

Table 12

Allocation of ration to growth and reproduction with age in the anchovy

Age (yr)	Weight (g)	Percent of Ration Growth	Reproduction	Ration (cal)
1-2	7.33-13.25	5	8	127
2-3	13.25-19.33	3	10	214
3-4	19.33-24.69	2	11	312

The growing proportion of food diverted to reproduction is generally associated of course, with the reduction in growth rate for most species approaching maturity. Large quantities of research data have been summarized by Brett and Grove (1979), Brett (1979) and Ricker (1979) on growth and physiological energetics in the laboratory and in controlled habitats (e.g., aquaculture). The "average observed" energy budget of fishes shown by Brett and Groves (Fig. 18, p. 337, 1979) is summarised as follows:

Table 13

Indicative values for marine fish of partition of energy ingested

	Remaining	Loss
Gross Energy Ingested	100%	
Faeces		20%
Digestible Energy	80%	
Non-faecal losses		7%
Metabolisable Energy	73%	
Lost as heat		14%
Net Available Energy	59%	
For Metabolic Maintenance		7%
Lost as heat		30%
Net available energy for Growth and Activity (the Conversion Efficiency):	22 percent	

Note that growth and activity components of metabolism are on the same line, since they are completely interrelated: (activity level determines growth for a given physical milieu and level of food consumption). The actual figure for efficiency, as noted elsewhere in this document, depends greatly on body size and activity level. Jones (1982) notes that efficiency of conversion of assimilated energy into body tissue (growth), is highest for juvenile stages, and decreases with size and age to close to zero for old fish. In general however, he suggests that for adult fish (of close to commercial size) the value of 10% due to Slobodkin (1961) may not be inappropriate.

Recent attempts to directly relate metabolism expenditures on activity to observed growth rates (Kerr, 1982) need to be re-examined in the light of what is now known on the relationship between swimming energetics and the non-uniform distribution of food resources. Whereas a fish with a passive feeding strategy (e.g., one held in the laboratory), responds positively to food presentation up to its satiation point, in nature continuously swimming fishes have to balance energy expended in moving between food patches, and to integrate the availability-vulnerability of their food resources in the environment as a whole (Vlymen, 1977; Sharp and Dotson, 1977; Sharp and Francis, 1976; Olson, 1982; Carey and Olson, 1982). Certainly, seasonal thermal and oxygen changes, long short-term variations in contagion and distribution of prey species, and the local density of competitor species determine the expenditure on activity, whereas growth and development of reproductive cells simply reflects the accumulation of energy above and beyond that required for metabolic maintenance and foraging. As a general evolutionary strategy, under conditions of stress, fish will also trade off energy used for somatic growth and that available for reproduction.

In nature, physiological parameters such as the rates of respiration, excretion, reproduction and growth in nature can only be taken as constants under prespecified conditions. To paraphrase Ricker (1979), it is unlikely that simple relationships exist between ingestion rates and subsequent growth or production for aquatic species. Data on growth should be collected and portrayed in the context of ambient conditions: temperature being the primary controlling variable. However, as we have just noted, the spatial distributions and abundances of food items and the work done by an organism to obtain them, are also important.

Although many general considerations have been summarized regarding costs of locomotion, particularly in fishes (Hoar and Randall, 1978; Webb, 1975), it is still an unusual paper which directly estimates "work done" as a variable in a close to natural context, rather than using literature or laboratory-derived estimates of these extremely variable energetic expenditures. Much more effort will be required before these variables can be realistically quantified, and what is intended in this chapter, is principally to provide an introduction to these important yet difficult-to-quantify, concepts.

Many, or even most organisms in the marine environment start life in the plankton, and subsequently make the transition with size into higher trophic levels. Although growth and capacity for persistence are overriding objectives among living organisms, at small time and distance scales, superior efficiency is needed to satisfy these energetic requirements, while being sufficiently abundant in a risky environment to be individually expendable.

The general relationship between size at reproduction and turnover time in the marine environment is a clear indication that despite wide variations in body size and form, behaviour, mobility and food preferences, certain energy limitations exist at each scale in order to ensure an organism's persistence and colonisation potential.

Developing from simple fission reproduction, and proceeding to encapsulation of larger embryos or live-bearing strategies, the evolutionary motivation for complexities in reproduction and life history, seems to be one of investing as much energy as necessary in order to place fertilized eggs into a "safe" nurturing environment (Alvarino, 1980; Sharp, 1980a; 1980b). Paraphrasing the chicken-egg truism, this is equivalent to defining an organism and its life strategy, as an egg's way of ensuring survival of another egg! In practical terms, reproductive strategies and adaptations appear to be geared toward optimizing either the numbers of eggs spawned into the immediate environment, or optimizing the selection of those larval or post-larval stages and life history characteristics which would give the best chance for the gametes to be placed into a less uncertain, more surely nurturing, situation. The elimination of (individually) high risk, by internal fertilization, is a solution which stabilises the environment of the embryo, and permits the young to enter the ecosystem at larger, apparently more secure sizes. The nurturing of young in the case of species exhibiting parental care, shifts the emphasis of natural selection from the developmental problems of short time and distance scales, i.e., those pertaining to the distribution of the eggs and larvae, to those pertaining to the adults. This would suggest that the energy spent per individual adult fish, might well be less in the highly evolved livebearing fishes or those exhibiting parental care than those producing millions of eggs and few survivors. However, numbers of young produced by live bearers are less numerous, (K-strategists) and this limits the rate of colonization of vacant environments. This is in sharp contrast to the situation for the highly fecund species with pelagic eggs (r-strategists). These differences in capabilities of pelagic larvae and liveborn young certainly sets the stage for very different environmental requirements of juveniles, as well as conditioning the very different selection pressures involved.

The oviparous species' eggs, like phytoplankton, are carried about by turbulence unless they are spawned on substrates, either floating, (as for flying fish Hirundichthys), or laid on bottom, e.g., herring (Clupea species), or are demersal, i.e., more dense than sea water, and sink close to the bottom where turbulence plays a more subtle role in their development. These latter types of eggs are from about 0.5 mm to 3.5 mm in size range, with pelagic eggs being about 1 mm in diameter. The larvae hatch after various times, depending primarily on temperature (Theilacker and Dorsey, 1980), and have from a few days to about ten days after hatching in which to use their yolk for further development. After the yolk is finished, they have from a few days to a week to locate food. The timing of hatching in relation to the availability of appropriate food and abundance of predators, is obviously critical (eg., Alvarino, 1980).

Parrish, Nelson and Bakun (1981) describe the zoogeographic discontinuities within the California current, leading to the various pelagic habitats, each with its local transport characteristics. Where offshore transport dominates, the major fish species do not have epipelagic eggs, but live bearers (e.g., *Sebastes*, and embiotocid species) are abundant, and species producing demersal eggs predominate. Wherever closed gyral systems occur, pelagic fishes with epipelagic eggs and larvae are found. Other, more migratory pelagic species appear seasonally, entering the coastal system in search of appropriate nursery areas. They then spawn, and depart. As seen from each organisms's own perspective of its habitat, the problems of transport of energy and materials into and out of local situations is one of mobility, or of scale of the habitat, for each of the various predators and their food components.

Plankton communities vary tremendously in relative motility, with motile forms ranging over kilometers in the course of periods of a week or less, but the basic dispersion of these groups is still related to the physical transport processes of the water column occupied.

The less mobile marine forms, small invertebrates, fish eggs and larvae and their forage, are known to carry on their interactions on the meter-day scale (Sharp, 1980a) whereas the more mobile post-transformation fishes explore kilometers over periods of days to weeks. Large, more widely-migratory (exploratory) species range over oceans on time scales of years, grazing down "local" abundances, i.e., smoothing out patches and adding or removing energy and materials, which are then transported or dissipated over very great distances. Each of these situations requires consideration when sampling the population or interpreting sample data from field or laboratory experiments.

15. DIVERSITY AND STABILITY OF FISHERY ECOSYSTEMS: ARTEFACTS OF SCALE?

Considerable progress has been made recently in our understanding of marine systems, and the reader is referred to texts such as Longhurst (1981), (Platt, Mann and Ulanowicz, 1981) for a description of recent thinking in biological oceanography. Some broad outlines and mechanisms underlying biological production in the ocean are evident, but as yet, the ability to make predictive statements on how much, of this production and in what form, is translated into fishery yield, is still limited. From an examination of the available information on landing trends for those fisheries with significantly long periods of historical catch data, Caddy and Gulland (1983) distinguished empirically between four classes of fish stocks:

(1) those for which steady-state assumptions seem to have applied for a significant period of time;

(2) cyclical fisheries characterized by fluctuations in landing of an apparently regular periodicity;

(3) those showing strong irregularities in landings from year to year; and

(4) fisheries characterised as spasmodic or transient. Obviously this classification is empirical, in that landing trends are affected both by "pulses" in fishing effort (Caddy, 1983) and by environmental factors *per se*, acting through fluctuations in recruitment success from year to year (see, eg., Holden, 1978 who documents these for one of the more stable marine areas, the North Sea).

The main point of interest here however, is that viewed in retrospect, the fisheries statistics of this century suggest that the species (and areas) where steady state assumptions appear to be valid are relatively limited in number extent and duration. Conversely, areas of high but irregular fish yields (eg., the upwelling off Peru, eg. Rasmusson, 1984) and the Kuroshio current, (Kondo, 1980, Kawasaki, 1980) are among the most productive of fishery areas. Strong long-term variations in hydrographic conditions, and hence in production levels and dominant species seem to be the norm, and the organisms present have accomodated their life cycles to this situation (Table 10). Needless to say, human fishing pressure and variations in exploitation patterns (e.g., Figure 61), can also significantly influence the species composition present even in these areas, but should be placed in their appropriate context of a variable 'carrying capacity' of the environment for the species in question.

The point can also be made that resource assessments, largely from a lack of suitable data over a long enough time period, have tended to make simplifying "equilibrium assumptions "over the medium-term period of 10-12 years. Multispecies interactions, and their perturbation by fishing and by processes beyond human control, are other factors that suggest caution be exerted in the

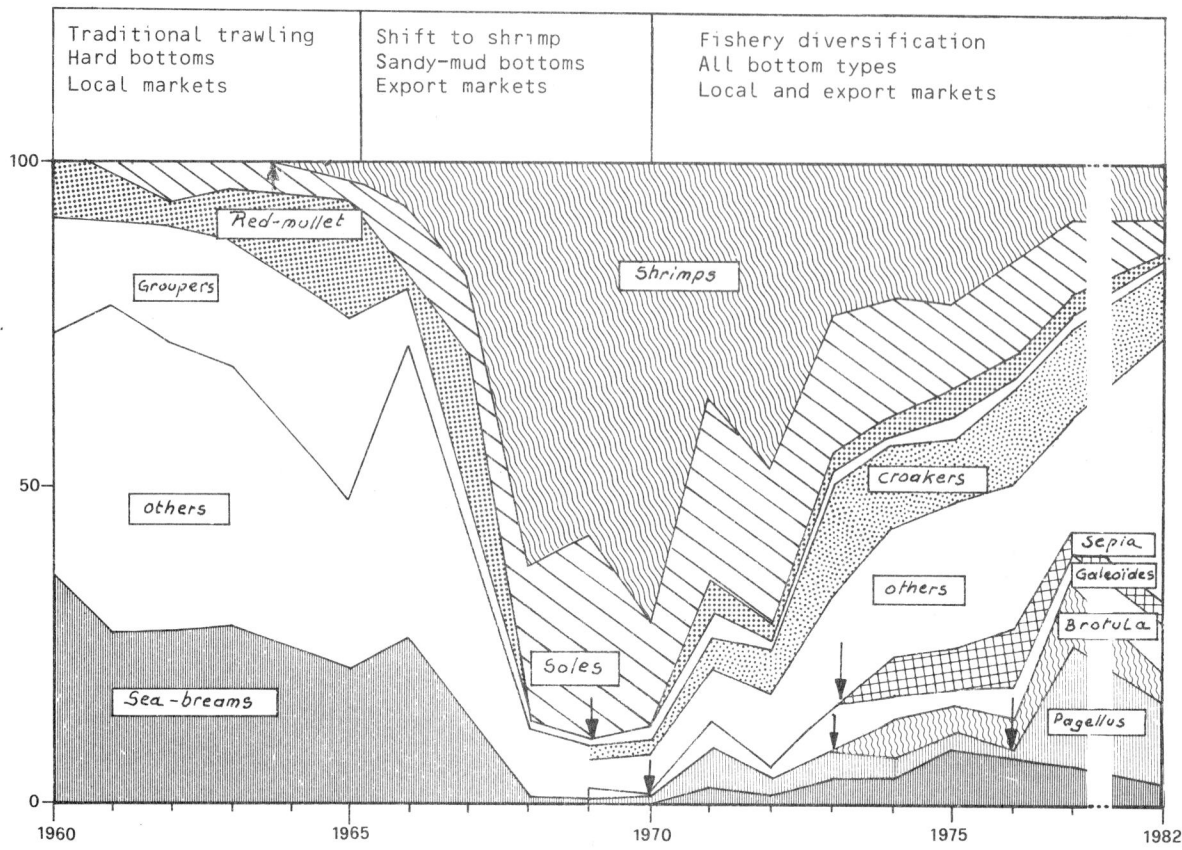

Figure 61 Changes in percentage species composition of landings in the Senegalese trawl fishery. The arrows indicate the appearance of new targets in the fishery (From Gulland and Garcia, 1984)

uncritical application of equilibrium assumptions in single species stock evaluations, before the ecological and environmental context of the species in question has been considered.

In general, the relative 'success' of each generation of fish is determined by how well it satisfied a hierarchic series of the survival requirements characterizing each life history stage. The situation is complex, and Bakun et al.(1980) refer to the "just right" period as a "survival window". In high latitudes the possibilities for massive completion of early life histories are framed by the short seasonal production cycle; the spring bloom. Nearly all species schedule the time of their periods of reproduction in relation to this bloom, and many species have homing mechanisms to help them locate ideal spawning conditions or locations. One result of this is that relatively few species have evolved to take advantage of these short term, localized constraints compared with the number of species found in tropical latitudes. This imposes no absolute limit to the potential productivity of high latitude systems however, since the lower ambient temperatures, less complex food webs, and longer-lived species with limited migrations tend to make high latitude systems ecologically more efficient than tropical systems, i.e., more biomass is produced per unit energy turned over.

The enormous productivity of for example, the Peru Current prior to the decline in the Peruvian anchoveta, can likely be attributed to the large size of the upwelling system, but also to the anchoveta's lower position in the food web compared to, for example, herrings or sardines. Without abundant anchoveta to consume the productivity of the Peru Current, the sardine bloom in Peru and Chile in the early 1980's only reached a small fraction of the biomass which the anchoveta attained in previous years. This can be explained in part as a consequence of the different positions each species holds in the food web, as well as by differences in ecological efficiency resulting from the different swimming modes employed by anchoveta and sardines respectively (Sharp, in press; Weihs, 1973; Ware, 1975, 1980). Ursin (1982) has suggested that since the decline of the anchoveta stock, there is reason to expect that invertebrate herbivores have taken up the "slack" left by the abscence of anchoveta, resulting in less energy available for commercially exploitable fishes. Both mechanisms seem likely to be acting together, and the above change in species dominance may reflect yet another variable: the poleward shift of the preferred habitat of small pelagics. (It is interesting to note, however, that by 1986, after a period of low effort exerted on anchoveta, during which more favourable conditions for population growth of this species were evident, there has been a significant recovery of the anchoveta stock.)

Until recently, preconceptions as to the stability of tropical multispecies systems have rarely been questioned, except in the context of the effects of overfishing. There are however several examples in upwelling areas of small pelagic fish (eg., Opisthonema and Cetengraulis) undergoing remarkable fluctuations in abundance off Central and South America and in Baja California under little or no fishing pressure, and similar observations have been made elsewhere (Troadec, Clark and Gulland, 1980). Spectacular changes in species dominance such as the explosion of Balistes stocks in the Gulf of Guinea (Figure 62) for example, may be related in part to overfishing of Sardinella, but other environmental factors also seem relevant.

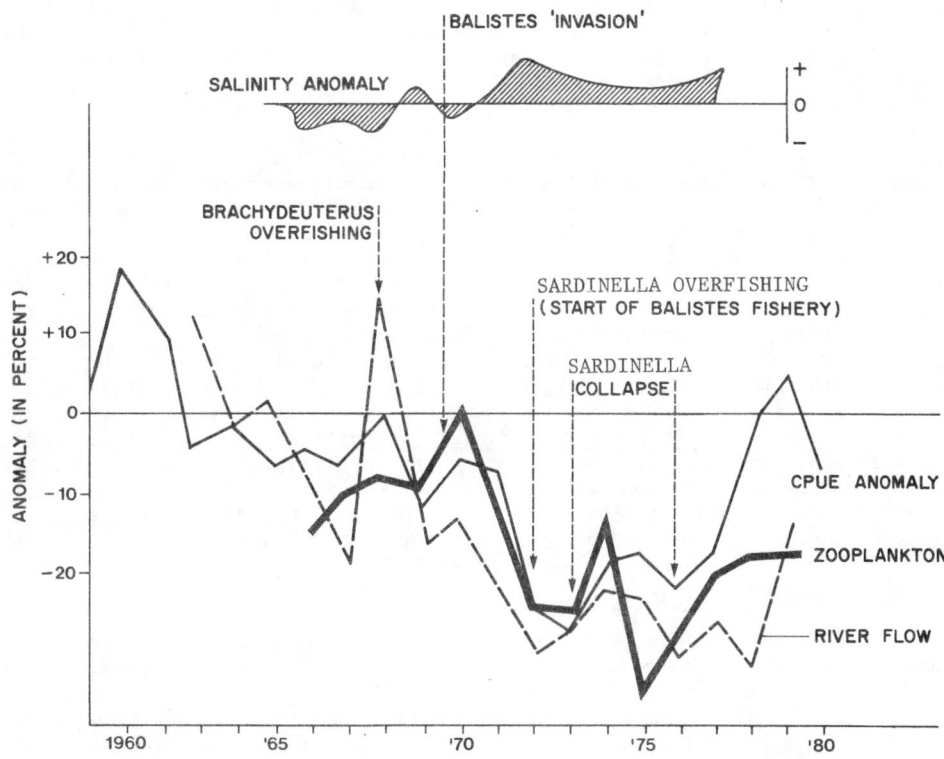

Figure 62 Trends in anomalies in the trawl fishery, river output, plankton abundance, salinity anomalies, and fishery events in the Côte d'Ivoire in relation to the changes in species composition of the resource (Sardinella collapse, Balistes eruption) (From Gulland and Garcia, 1984)

In the case of coastal or shelf fisheries in the tropics, such as those in the South East Pacific, the numbers of species taken in any given location may be relatively large in contrast to the higher latitude coastal or shelf fisheries, but the variations between contiguous areas are also likely to be considerable in tropical areas due to the relatively more numerous species, and their smaller, mosaic-like distributions. The integration of catch for coastal species statistics over a large area would tend to smooth out these variations, and obscure the scale and patterns of local breeding populations, thus arguing for tropical coastal resources, for a high degree of regionalization of resource management.

Ecological measures of diversity and stability are usually defined for relatively small areas in terrestrial contexts also: with units of pattern having dimensions of metres to kilometers for e.g., grassland biomes, as opposed to tens to hundreds of kilometers for most high latitude fisheries management units (see review in Pielou 1969 for theory). Enumeration of species and population descriptions are ideally carried out over homogeneous areas, and the portrayal of changes, or differences, should ideally be associated with habitat discontinuities. The major problems posed by the 'remote sampling' nature of most fisheries investigations (e.g., the "integration" of two or more fish communities in a single bottom trawl haul over several miles), makes this ideal sampling approach difficult to attain, but still important to have in mind when attempting to analyse results of such surveys.

In planning fisheries field investigations, insufficient weight may be given to financing overall survey and identification exercises, even though such a preliminary "mapping" of habitats, communities and resources, and analysis of the resulting information has been suggested (Caddy and Garcia, in press), as essential to stock assessment as well as, coastal resource planning and pollution impact studies. Such detailed investigations of the structure of marine communities have usually been confined to near-shore marine environments where much of the transfer of the methods of terrestrial ecology to marine science has occured. These investigations, especially of reef or outcrop areas (epifaunal or 'live bottom' communities), have suggested that the "fine structure" of bottom communities on the shallow shelf is of the same order of scale to terrestrial communities, as anyone who has swum over a coral reef can testify. This is not the case for level bottom infaunal communities of the shelf and deep ocean (see e.g., Thorson, 1957), although here also, ecological units of pattern are probably smaller than is apparent at first sight.

Measurements of stability and diversity are thus as sensitive to scale (time and space) as is their conceptualization. A study of communities with numerous migrant species will show stability when sampling is integrated over a year, whereas they will appear far less stable on the basis of monthly sampling. Choosing geographic coordinates in lieu of habitat characteristics for statistical boundaries can have a dramatic impact on the characterization of marine situations. The 1982-83 El Niño phenomenon has had marked affects on the fauna of coastal equatorial islands in the Pacific, with some species disappearing from atolls and reef communities completely due to changes in the wind field and subsequent current shifts. The coastal transition zones delimiting the tropical and subtropical waters had also shifted steadily poleward since about 1970 in the eastern Pacific (Sharp and Csirke, 1983). These changes would have confounded any geographically-based sampling or management system if the question was whether or not to limit catches of particular species in the transition zones. While stability-diversity studies on both Pacific islands, and at coastal sites on continental land masses, tend to show a phased disappearance of some species, with changing stability and diversity indices; sampling by large, fixed geographical units would not have told us the real tale of species displacement. Some stocks would have seemed to decline, and new stocks and species would have unexpectedly become abundant in other statistical areas, but movement of species boundaries would have been less evident. Such phenomena seem also to have occurred in West Africa in recent years.

Examples of major changes in abundance, distribution and species composition for neritic fisheries throughout the world are given in Sharp and Csirke (1983). Even though single fisheries tend to aggregate more than one stock and species into exploitation units which may or may not be single breeding populations, the instabilities of many neritic populations are quite evident from fishery data, both for so-called demersals and for pelagics. As noted earlier, the continuously stable population is a relatively rare case over historical time scales, and therefore ecological diversity will also be subject to changes in environments where unstable populations are fished. The aggregation of resource populations due to the mobility of the fishing fleet often obscures the true scale of spatio-temporal variability of these populations, and requires that fisheries data bases classify landings by area/time of origin rather than (the easier option), by boat and port of landing.

As we have suggested, the single-year life cycles of many species living on tropical shelf areas set the scene for very high fluctuations in abundance, and these populations share some features of the "bloom and decay" characteristics-typical of shorter-lived planktonic forms at high latitudes. This feature is in part due to the lack of correspondence between catch composition or relative species abundance of fisheries separated by as little as fifty nautical miles; (e.g., in the Indonesian archipelago; Pauly, personal communication; Bakun et al., 1980). If this is indeed the case, then it is likely that what we may be encountering at any one time is the result of differentially successful "local colonizations" by different species from an array of opportunistic species operating in an uncertain spawning context. The magnitude of change in dominant species abundance from year to year, and location to location, can often be considerable. Whether the specific early life history requirements are met or not, the presence or absence of predators, and the impact of locally intensive harvesting or closed seasons for inshore fishing, all set hurdles at which the larvae, juveniles or adults can be removed from the system, or alternatively allowed the time to grow and reproduce. What appears as relative stability in catch statistics at a national or regional level, may simply be the result of over-integrating the picture, i.e., looking at too great a part of the picture at any one time. This may be inevitable or even desireable from a national resource economics perspective but will lead to problems if not taken into account in deciding on the 'appropriate scale' of marine resource investigations, or more practically, in guiding the management authority as to the appropriate field of action of management measures for resources of a given bay, gulf, estuary, etc.

The above considerations can also be extended to include open ocean species such as the tunas. In the northwestern and eastern Pacific Ocean there has been a continual estimation of recruitment from yearly statistics of catch by age, but little subregional stratification has been officially reported. The impression generated from this large scale integration of catch data, is that tunas do not exhibit population fluctuations of the same order as other pelagic species; i.e., there are only 2-3 times changes in biomass or recruitment over the entire eastern Pacific yellowfin fishery, or over the skipjack fishery in the northwest Pacific.

However, in the eastern Pacific Ocean yellowfin tuna fishery there are regions which appear to support primarily "local" populations in some years, and mixtures of various independent cohorts which consist of both local and nomadic yellowfin from adjacent areas, in others. In Table 14 from Sharp and Francis (1976), there is a description of annual variations in cohort catches within three sub-regional areas. These can be considered to include "local" plus nomadic components of the eastern Pacific Ocean yellowfin population. The largest ratio between catches from any two semestral cohorts (fish recruited in the same 6-month period; any years) within any of the three

Table 14

For the years 1964-71 the data are presented for catch in short tons by semestral cohort in the three areas (N,5,S) within the CYRA (Commission Yellowfin Regulation Area). Also given are the percent of the total catch ($S_A + S_B$ + Big) by cohort within the areas. The category "Big" represents fish over 145 cm which we felt could not be aged under the present system. The percent of the individual semestral cohorts (S_A or S_B) caught in the three areas is also given. Note the erratic shifting of the dominant cohorts, semester A or semester B, in the catch as well as the shifting distributions of these cohorts in and among years

Year	North A	5 A	South A	Total A	North B	5 B	South B	Total B	Total A + B	Big
1964	27 452	9 401	5 209	42 062	33 561	5 881	17 515	56 957	99 019	2 921
% total A + B	26.9	9.2	5.1	41.2	32.9	5.8	17.2	55.9		2.9
% total A or B	65.3	22.4	12.4		58.9	10.3	30.8			
1965	18 967	13 512	6 406	38 885	24 064	14 164	8 386	46 614	85 499	4 543
	21.1	15.0	7.1	43.2	26.7	15.7	9.3	51.8		5.0
	48.8	34.7	16.5		51.6	30.4	18.0			
1966	7 769	23 128	20 176	51 073	0 292	11 394	14 771	36 457	87 530	3 626
	8.5	25.4	22.1	56.0	11.3	12.5	16.2	40.0		4.0
	15.2	45.3	39.5		28.2	31.3	40.5			
1967	20 699	9 564	7 664	37 927	29 482	8 572	11 867	49 921	87 848	1 802
	23.1	10.7	8.5	42.3	32.9	9.6	13.2	55.7		2.0
	54.6	25.2	20.2		59.1	17.2	23.8			
1968	16 361	23 921	13 552	53 834	33 917	22 132	3 128	59 177	113 011	1 602
	14.3	20.9	11.8	47.0	29.6	19.3	2.7	51.6		1.4
	30.4	44.4	25.2		57.3	37.4	5.3			
1969	22 437	20 034	9 030	51 501	34 887	29 587	5 648	70 122	121 623	4 888
	17.7	15.8	7.1	40.7	27.6	23.4	4.5	55.4		3.9
	43.6	38.9	17.5		49.8	42.2	6.1			
1970	39 197	15 942	10 529	65 668	43 476	13 257	11 125	67 858	133 526	9 176
	27.5	11.2	7.3	46.0	30.5	9.3	7.8	47.6		6.4
	59.7	24.3	16.0		64.1	19.5	16.4			
1971	12 372	18 719	14 453	45 544	17 357	25 283	15 712	58 352	103 896	9 277
	10.9	16.5	12.8	40.2	15.3	22.3	13.9	51.6		8.2
	27.2	41.1	31.7		29.7	43.3	26.9			

sub-regions, is about 5X, whereas between sub-regions for the same recruitment period the variation approaches 19X. The catches of yellowfin tuna taken together along the entire coastal zone of the eastern Pacific have been relatively stable since the full development of exploitation of this zone in 1961. Considering the catch of yellowfin within 200 miles of the coast from 1961 to 1976. The ratio of greatest to smallest catch is less than 3X. This shows the regional stability characteristic of populations with large nomadic components when integrated over large areas, in contrast to local or sub-regional variations.

Table 15

Trajectories of catch trends from 1970-77 for some key world pelagic resources

Species	Area	(A) Peak Catch	(B) Low	(A/B) Catch Ratio	
Caranx hippos	West Africa	28 221	1 036	27.	+
Orcynopsis unicolor	West Africa	2 600	100	26.	−
Trachurus capensis	Southwest Africa	690 164	62 300	11.	+
Trichiurus lepturus	Southwest Africa	28 545	3 800	7.5	+
Trachurus trecae	Southwest Africa	273 700	31 298	8.7	−
Sardinella spp.	Southwest Africa	142 200	20 986	6.8	− +
Scomber japonicus	Peru	65 000	8 700	7.5	+
Scomber japonicus	Northeast Atlantic	39 000	6 262	6.2	−
Rastrelliger spp.	Eastern Indian Ocean	16 300	2 000	8.2	+
Rastrelliger kanagurta	Eastern Indian Ocean	203 100	35 403	5.7	
Anchovies	Western Indian Ocean	118 062	16 900	7.0	−
Psenopsis anomala	Northwest Pacific Ocean	13 000	1 994	7.0	−
Sardinops melanostictus	Northwest Pacific Ocean	1 420 512	16 900	84.	+
Engraulis mordax	Eastern Pacific Ocean	289 002	44 600	6.4	+
Cetengraulis mysticetus	Eastern Tropical Pacific	168 081	15 551	10.8	+
Trachurus symmetricus	Eastern Pacific Ocean	50 149	9 400	5.3	+
Sarda chiliensis	Southeastern Pacific Ocean	74 700	4 341	17.2	−
Scomberomorus sierra	Peru	2 279	400	5.7	+
Engraulis ringens	Peru	13 059 900	807 175	16.	−
Sardinops sagax	Peru-Chile	1 467 555	68 600	21.	+
Trachurus trachurus	Peru-Chile	839 805	111 300	7.6	+
Thyrsitops lepidopodes	Chile	7 200	630	11.6	−
Cetengraulis edentulus	Venezuela	496	850	5.8	− +
Decapterus russelli	Malaysia-Thailand	109 337	9 800	11.2	+
Scomberoides spp.	Indonesia-Philippines	5 186	500	10.	+

Plus and minus signs in the Table represent directions of trends during the reference period. The indication − + implies sharp changes in both directions of the order indicated

Examples of pelagic fisheries which have had annual catch peaks of more than 5X the lowest catches during the period 1970 to 1977 are tabled above (Table 15). It is expected that some of these catch variations are due to changes other than those interactions believed to be environmentally induced (i.e., which are due to changes in fishing effort, or to unreported catches). However, many of these fluctuations reflect real recruitment and consequent biomass fluctuations, which stem to some degree from local habitat variation; particularly large increases in catch during this period.

Paloheimo and Regier (1982) have pointed out the different considerations of scale which would be relevant and appropriate in approaching the multispecies fisheries of, for example, the Great Lakes and the Grand Banks of Newfoundland. They suggest that in order to cope with multispecies-ecosystem related fisheries, research will have to adopt empirical approaches if it is going to be of practical value to fisheries management. Highly theoretical approaches are only of relevance when either very little is known about a system, or very much is known. They also point out that many management related studies tend to be vague about what has happened as a result of previous management actions, or what to expect from them in the future. Although fisheries can induce dramatic changes, the major causes of large scale system effects are still variation in abiotic properties of the ecosystem. The biotic changes will be responses to these. Figure 15 from Regier and Henderson (1973) is provided here to show that this way of classifying the numerous approaches used in fishery studies also implicitly explains the rather low precision characteristic of management-related research. It is also noted that this low precision is not all due to the unpredictability of fish populations but is due in part to the failure to consider appropriate environmental covariates that would partly explain the "apparent random noise" in multispecies fisheries.

As was also concluded in Sharp and Csirke (1983), Paloheimo and Regier (1982) observe that:

"Any attempt to explain changes in multispecies fisheries that ignore concomittant changes in the abiotic variables appear to be futile. Similarly, any modelling of the freshwater or marine system without incorporating the role of environmental variables also seems to be doomed to fail".

This emphasizes the importance of comparative fisheries research. Careful examination of biomass composition changes in local or regional fisheries often shows a relatively stable total biomass, with compensatory changes occurring among the various species components. This nearly constant condition is only likely to occur in most marine systems as long as the primary production remains relatively stable, and where the replacement species are ecologically equivalent.

The statistical data base we have on most industrial fisheries in the tropics is very brief in duration; dating from the 1960's or even the early 1970s in most cases. These data illustrate the short-term variations in catch from pelagic populations. What changes can be expected in the long-term will only become evident as the data series and experience with these fisheries accumulates.

The trends in regional catches and relative abundances of the oceanic opportunist species such as the tropical tunas, large billfishes and other cosmopolitan species, vary in the order of two to three times, but rarely as much as five times. Where records indicate changes of such magnitude for cosmopolitan oceanic species, they usually correspond to changes in economic factors or in effort distribution patterns, rather than representing relative abundance changes.

For demersal fish it has been noted that specific bottom type and depths may each have a characteristic fish fauna (e.g., Grady, 1971; Domain, 1972; Caddy and Iles, 1973) although here the concept of specific fish "communities" that have a tendency to share common population ranges appears less convincing than for bottom infauna. The use of the term "fish community" tends now to be more commonly applied to specifically recognizable "live bottom" areas, e.g., coral reefs, eel grass or kelp beds. Even here it may be regarded as being broadly descriptive, and rather difficult to verify statistically since, for example, coral reefs contain a number of different habitats, each with its group of more or less characteristic species.

As noted by Parrish and Zimmermann (1977), reef fish can themselves be classified into four categories: Reef restricted species that only occur on reefs, reef related species that spend at least some but not all of their time on or around reefs, reef indifferent species that may or may not spend some time around reefs, and non-reef or occasional species that are atypically found there.

More generally, one may first classify species into residents and transients – the latter, often important, species only occurring in a "community" on a seasonal basis (e.g., Tyler, 1971). Even for benthic invertebrates, it is clear that with successive sampling, the total number of species encountered goes up in a generally assymptotic fashion with number of samples taken (Figure 63), so that to define the number of species in an area for calculation of diversity indices, etc., it will be necessary to define the number of samples taken.

16. SAMPLING THE MARINE ECOSYSTEM: THE CONCEPTS OF CONTAGION, COMMUNITY AND ASSEMBLAGE

Sampling and Fishing: Methods and Constraints

Basic information on the numbers and species of fish populating a given marine area is obtained for the most part indirectly, that is by inference from fish catches, in particular by fishing gear ("remote sampling devices") of various types, each with its own specific efficiency and selectivity for species and sizes. More "direct" observation techniques such as sonar, underwater photography, and observations from vehicles such as submersibles or by scuba, have been used in some areas for quantitative estimation of fish and shellfish populations, but also have their particular difficulties and limitations, being susceptible to problems in identification and calibration, and to the effects of the observer on behaviour of the fish being observed. For visual methods, the general limitation is also on the area of action of the sampling device, given that although average fish densities per area of sea surface may be in the order of 1-10 fish/m^2 overlying dense schools, they may average 1 fish per 10-1 000 m^2 over the whole fishing ground, and of course much lower "average" densities are the case for large high seas predators such as tunas.

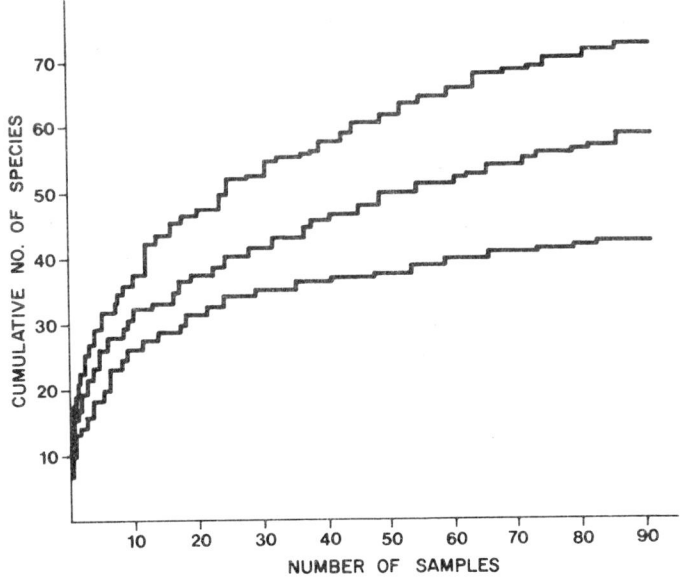

Figure 63 Illustration of how species numbers approach an assymptote with sample frequency or area sampled (Redrawn after Smith and Tyler, 1973)

It is important to remember however, that the effectiveness or "fishing power" of vessels sampling a fish population are not just functions of gear design: a skipper who has special knowledge of areas of higher density has a competitive advantage and in general, predators take advantage of patchiness of prey in this way to increase their "benefit-cost ratio". Thus, a shipper's knowledge is a significant component of his vessel's fishing power, which can be at least as important in practical terms as the vessel engine horse power in determining the catch rate of those species available to the gear.

This second quantity, "availability" or "vulnerability" to fishing is again, a factor that determines the possibility of sampling or commercial fishing with a given vessel/gear type, and may change from time to time. The high vulnerability of herring, cuttlefish, grouper and many other species to fishing while spawning, is of course a factor that a manager needs to take into account in deciding on fishing seasons, closed areas etc., not only in designing sampling programmes.

Distribution Patterns

Sharp and Francis (1976) estimated the average density of yellowfin tuna larger than 40 cm in fork length in the eastern Pacific to be 34.5 mg/m^2. Of course tunas of 40 cm or more weigh considerably more than 34.5 mg, and also occur in schools. Sharp (1978) estimated that the typical tuna school sought by modern fishing gear is a composite of "core" schools, i.e., groups of fish of similar size and similar or unique history, comprising 2 000 to 3 000 kg units. The fishery in the eastern Pacific samples schools which are greater than some threshold size (specific to each fishing captain's criteria). In examining several localized series of data for insights into environmental effects on school size, Sharp (1978) noted that in two periods which were characterized by different thermal profile regimes, the proportions of different size schools captured changed dramatically. The following table shows these results:

The observed change is in the decreased numbers of small sets (0-10 tons) made and in the proportions of sets containing more than 20 tons. Of the 881 sets made from 1 March - 1 April, only 9 percent were greater than 20 tons; in the second period there were 499 sets and 22 percent were larger than 20 tons.

The point Sharp (1978) makes is that school sizes shift with suitable habitat volume, hence local "core school" density and interaction rates change. During the latter period referred to above, the habitat volume was less than that for the earlier period, hence more core schools met and aggregated. This example causes one to consider that in sampling fish stocks, system dynamics, as well as availability and selection by the samplers, in this case the fishing captains, must all be considered.

Table 16

Percent total successful purse seine sets in class

Short Tons =	0-10	20	30	40	50	60	70	80	100	>100
Mar 1 - Apr 15	71	21	4	2	1	1				
Apr 16 - May 30	59	20	8	4	2	2	2	1	1	2

As a simplistic example, again looking at the eastern Pacific yellowfin tuna in its context as a dominant nomadic predator in this large system, consider that the 2-3 ton core school represents about 50/80 km^2 of ocean surface, and that a not unusual 200 ton aggregation comprises the entire expected biomass of about one half a 10° longitude by 10° latitude area of the eastern tropical Pacific. In these schools there is an inverse relation between average fish size and its probability of occurrence in larger aggregations. Consider a 200 ton school of 50 cm fish, consuming at least 10 percent per day biomass equivalents, in order to sustain itself: at normally observed densities of coastal resources, such predator removals of 20 tons per day would represent a significant perturbation of the local fish resource and food webs.

What this illustrates again is that fish are not randomly but contagiously dispersed in their habitat; the degree of contagion to a large extent being dependent on both the life history stage and on the size of the unit area sampled. The statistical distribution that perhaps provides the best general description of the probability of a given number of organisms being captured in a unit sample is the negative binomial distribution (Figure 64), and this has also been found to provide a good description of the number of fish in successive samples taken with an otter trawl (e.g., Taylor, 1953)[1]/, and well describes the distribution of benthic populations also (see Elliott, 1971). Sampling fish with "swept area" type of gear (trawls, dredges) poses special problems in interpretation of data, even if these may be less pronounced than for capture methods that depend on fish attraction (traps, baited hooks) or for pelagic fishing gear (gill nets, seines and weirs) where an adequate sampling theory has, by and large, yet to be developed. The problem with all fishing gears is basically one of defining their selectivity and the availability of different sizes and species to the gear, before allowing for it in assessment of the true abundance of fish; bearing in mind that most successful fishing gears have been developed (often over long periods) to provide the maximum differentiation between desirable and undesirable sizes and species.

Sampling demersal communities

The bottom trawl is probably the best understood type of fishing gear, and also perhaps the gear that for demersal fish comes closest to satisfying the main sampling requirement, namely, that catch rate for a given species in an area is usually roughly proportional to abundance. However, even in this case, species differences in behaviour, e.g., in response to herding by the warps, and differing effectiveness of escape responses, together with mesh size, all dictate the species composition of the catch. Trawling characteristics such as direction relative to the tide, and speed of tow, headline height, and presence of rollers or covers, also affect the catch composition. Perhaps, however, the most important of all factors in relation to a study of demersal fish "communities" (defined as those groups of species that generally occur together in the same habitat) is the duration of the tow, since for long tows (over several kilometers), the trawl may be artificially "blending" several groups of organisms, each originally occupying a distinct subarea of sea floor within the total path swept by the trawl.

As noted by Thorson (1957), level (theoretically "trawlable")[2]/ bottoms, make up at least 90 percent of the sea floor of the world's oceans and are inhabited by a mosaic of distinct groups of bottom-dwelling invertebrates. Thorson suggested that each group of organisms is made up of a relatively fixed number of species, making up a "community": the "communities" differing with sediment type and depth.

As noted already one may first classify species into residents and transients - the latter, often important, species only occurring in a "community" on a seasonal basis (e.g., Tyler, 1971).

---

[1]/ It has also proved a useful description of several other important statistical distributions in fisheries science; e.g., the probability distribution for the size of annual recruitments for a single species population (see, e.g., Hennemuth, Palmer and Brown, 1980)

[2]/ Most of these level bottom areas are of course abyssal and not trawlable!

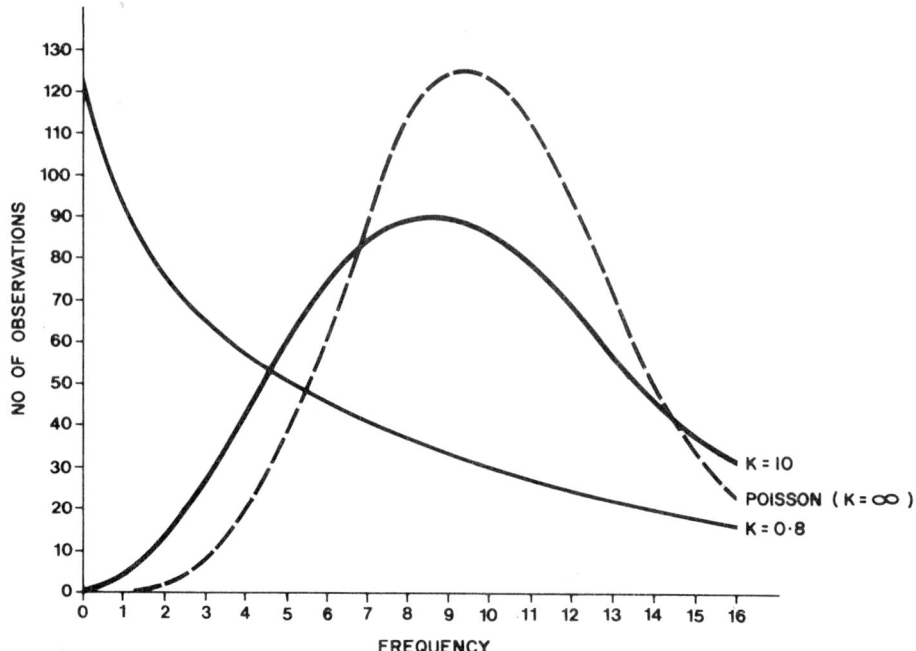

Figure 64 The negative binomial distributions compared to the limiting form, the Poisson. Each distribution has a mean of 10 and consists of 1 000 observations (From Taylor, 1953)

Even for benthic invertebrates, it is clear that with successive sampling, the total number of species encountered goes up in a generally assymptotic fashion with number of samples taken (Fig. 63), so that to define the number of species in an area for calculation of diversity indices, etc., it will be necessary to define the number of samples taken.

For this sort of reason, and because, unlike the term community, it leads to field studies of a more objective kind, the concept of a fish "assemblage" has been coming more into favour in recent years. Based originally on work by terrestrial botanists (e.g., Whittaker, 1967), the concept of gradient analysis became accepted, whereby apparent changes in flora (and of course associated fauna) were explained in terms of a common response to gradients in environmental parameters (e.g., altitude, pH, temperature, drainage, etc.). As a consequence, a group of species typically occurring together were considered to be responding to a common environmental variable, rather than from any intrinsic "attraction" between them. Many exceptions may be found of course, e.g., those that fall under the terms parasitism, commensalism, symbiosis, and obligate predation, where two or more species co-occur independently of the environment so long as this is not wholly unfavourable. The gradient concept and the associated idea of species assemblages (i.e., those organisms that due to independent but common responses and environmental preferences tend to occur together) has a lot of attraction, especially since unlike the community concept, it allows statistical testing of the underlying hypotheses, and encourages study of the physiological ecology of marine organisms.

The fish assemblage may in practice be considered an alternative form of classification to that discussed elsewhere in this paper, namely the food web, and it may be worthwhile spelling out some of the differences, while noting that the two approaches are strictly complementary and should together logically underlie any approach aimed at understanding multispecies fisheries.

The fish assemblage consists of those species available to sampling in an area, which from statistical analysis can be shown to occur together on a recurrent basis. Not all of these species will necessarily be linked trophically, although many will be (either as predator-prey, or by sharing common predators or prey). In fact, since the components of an assemblage will reflect the method or methods of sampling used, they may (unlike food web components) tend to be of a relatively restricted range of individual sizes as adults.

Various statistical tools have been used to show the degree of co-occurrence of species in a series of samples, and a number of indices of similarity between different samples have been used (see, e.g., Greig-Smith, 1964; Whittaker, 1967; Day and Pearcy, 1968; Pielou, 1969). These can be resolved into groups of stations with similar fauna, or groups of species with similar spatial distributions, by using cluster analysis e.g., Romesburg, 1984 to show the hierarchical similarity of different groups of stations (e.g., Figure 65). Various routines exist (see, e.g., Davies, 1971) for arrangement of a matrix of similarity indices in a hierarchical fashion as a dendrogram or contingency table, if systematic sample data exist. Methodologies of treatment of multispecies sampling are progressing rapidly, and in particular, Principal Component Analysis is coming into widespread use (see Mahon et al., 1984). Even before such systematic approaches are tried, as noted earlier, separate maps can often be prepared for resident (as well as on a seasonal basis for transient species), to show the degree of overlap of different resources and their fisheries.

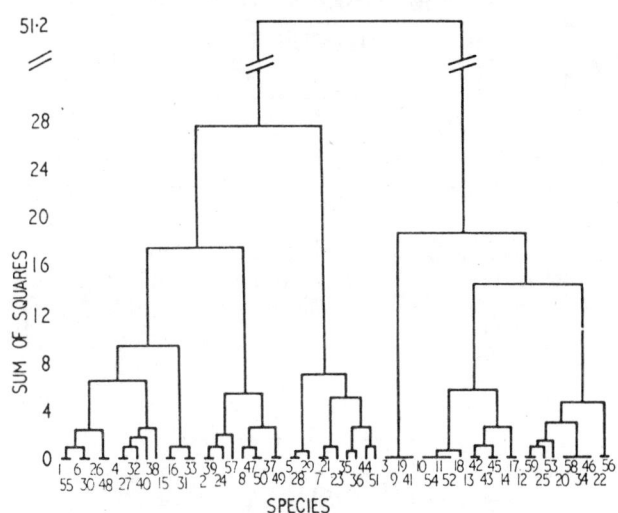

Figure 65 Dendrogram for hierarchical classification of stations (from Hughes and Thomas, 1971): the letters represent subgroups of stations

Characterizing fish assemblages in this way, even approximately, appears to be one necessary precondition to the understanding of multispecies stocks, and could usefully be accompanied by the approach to analysis of stomach contents, and the preparation of food webs in the preliminary form suggested in this paper.

As pointed out by Tyler, Gabriel and Overholtz, 1982, "Fishery agencies in many parts of the world already have enough distributional data to map assemblages ...(and)... there are already enough data already collected ... so that regulars may be distinguished from seasonals. There are also enough fish stomach analyses in many important fishery regions to construct first approximations of energy usage groups ...". These "energy usage groups" consist of the (one or more) trophically linked subset of "regulars" (year-round resident species) in an area, and Tyler suggests that these subsets form the natural multispecies management units for a fishery (e.g., Figure 66). This will require a particular approach to research and management of fish stocks namely, management by "Assemblage Production Units"(APU's), each consisting of several species. These demersal "energy user groups" are considered to be affected by three main driving functions: the fishery, the physical environment and its fluctuations, and the impacts of the transient species (as occasional predators, prey or competitors). This preliminary categorization thus goes a long way towards elucidating the likely interrelationships between important species. Although the concept of assemblages consisting of year-round residents, may be rather restrictive for (eg. estuarine and upwelling) areas where oceanographic conditions make year-round residence impossible, MacDonald et al. (1984) have discribed assemblages that retain their identify despite seasonal bathymetric displacements. There is also some evidence (Overholtz and Tyler, 1985) that assemblages will persist over significant period of time despite drastic changes in relative abundance of the constituent components.

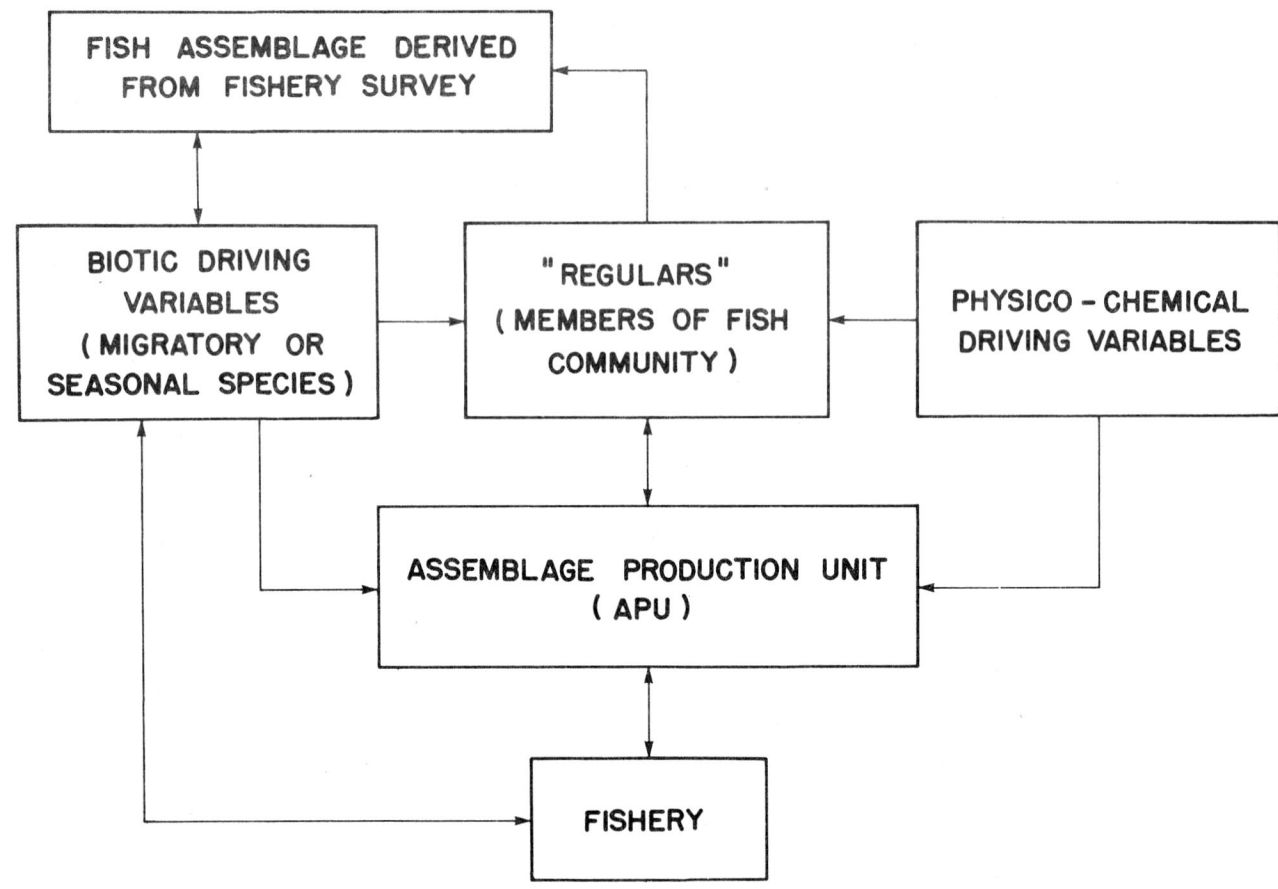

Figure 66 Schematic approach to research and management by fish stock assemblage (After Tyler, Gabriel and Overholtz, 1982)

We should also bear in mind (as shown in Figures 7 and 8) that a food web is a static representation of a recurring sequence of events in the sea. As implied by these illustrations, the biomass of a given component at the present time, should in theory if not in practise, be translatable backwards to the period of time when synthesis originally occurred, if we were to know the efficiencies of transfer of material between different components, and its rate of usage for metabolic purposes by marine organisms. This time sequence of events is not always clear, however, because of the existence of "loops" in the form of scavengers, detritivores, and by cannibalism, which delay the "upward" movement of material in the food web. Nonetheless, Figure 8B can be regarded as an analogue to single species cohort analysis, and is perhaps a visual equivalent to the multispecies cohort analysis proposed by Pope (1979).

For demersal and pelagic fish also, it has been noted that specific bottom type/depths/water masses may each have a characteristic fish fauna (e.g., Grady, 1971; Domain, 1972); and that fish aggregations are dynamic and occur in response to environmental gradients, but also in response to spawning and feeding, so that the concept of specific fish "communities" that have a tendency to share common population ranges appears less convincing as a general rule than for bottom infauna. The use of the term "fish community" tends to be more commonly applied to specifically recognizable "live bottom" areas, e.g., coral reefs, eel grass or kelp beds. Even here it may be regarded as being largely descriptive, since, for example, coral reefs contain a number of different habitats, each with more or less characteristic species.

17. THE EFFECTS OF TEMPERATURE AND BODY SIZE ON FEEDING AND ON THE NATURAL MORTALITY OF PREY

The relationship between feeding rate and growth has been a major preoccupation of marine biologists for the last 30 years or so, and a wide and varied literature exists (see Ivlev, 1961; Winberg, 1960 for early key references). Much less attention has been paid to the effects of feeding rate on the mortality rate of those species in the trophic level immediately below, that form the food of the species in question.

Conover (1978) summarized a representative set of data on feeding rates in relation to body size and temperature for a range of invertebrates and marine fish, and although he concluded that sweeping generalizations on the rate of feeding in large particle feeders macrovores are difficult to make if the whole data set is taken together, he arrived at two main conclusions:

(1) an increase in temperature increases the feeding rate;

(2) small fish require a proportionally larger ration (i.e., percent body weight consumed per day) under given conditions than large fish.

Looking at correlations between body size and food consumption, even within taxonomic groups, is unlikely to be a very fruitful exercise if the normal level of activity of a species is not taken into account. Since activity contributes greatly to metabolic rate and increases with temperature, inclusion of ambient temperature in any comparison improves our ability to estimate feeding rate. The majority of available data are from tank experiments whereas most swimming activity is in response to feeding, hence producing usually relatively low "maintenance" levels of activity and feeding rates, although the contrary tendancy of "feeding to saturation" characteristic of aquaculture operations for example, to some extent compensates for this.

Despite the above, there are obvious basic differences between the level of activity and hence feeding rate between for example, inactive and active perciformes, e.g., a flatfish and a mackerel, or between an octopus and a squid (both cephalopods), even at the same temperature and body size. This is seen clearly in Figure 67 where the feeding rate is plotted against body weight for the data from Tables 5-24 of Conover (1978), (all diets combined). The squids and mackerels have feeding rates that are at least two times higher than for the less active bottom dwellers, the flatfish and octopus. From preliminary examination of these data, it seems that there is relatively less variation in feeding rate for fish in a given ecological niche or activity pattern. Bearing this in mind, we looked more closely at the most extensive series of data that was available, namely for a range of predatory bottom-dwelling fish (flatfish, gadoids, groupers, etc.) shown as circles in Figure 67, feeding on a variety of diets which are not distinguished in this analysis.

Expressing this relationship empirically by a multiple regression, we can fit:

$$\log_e(r_{T,w}) = A + B \log_e(\bar{W}) + C(T)$$

A preliminary calculation of the parameters of this equation for a limited set of 59 data points taken from Table 5.25 of Conover (1978) (and ignoring the nature of the diet), was carried out. We find for 13 species of demersal fish ranging in size from 20-633 g under ambient temperatures of from 8.6-20°C, the following multiple regression of the above form:

$$\log_e(r_{T,w}) = 1.841 - 0.286 \log_e(\bar{W}) + 0.048 \bar{T} \ldots\ldots(A)$$

where $r_{T,w}$ is the feeding rate (% body weight per day) and $\bar{W}$ and $\bar{T}$ are mean (or in some cases, the median) body weight and temperature respectively. An R^2 of 0.55 suggests that the variables considered explain only about one half of the variation observed; and this relationship should be re-examined when a larger size range and bigger data set are available. Despite this, our preliminary analysis also suggests that of the two independent variables, body weight is the more significant, explaining almost 40 percent of the variation - a smaller percentage being explained by ambient temperature. Specific effects of activity pattern and type of diet probably account for a significant part of the remaining unexplained variation[1]/, but even if a quantitative data set were available or could be easily defined which included these unknowns, it seems unlikely that they would individually be more important than body size in determining feeding rate.

1/ The concept that the "resting" or maintenance level of metabolism and food consumption is roughly half that of a species in its natural environment is due to Winberg (1960), but does not take into account difference in specific activity level of different species

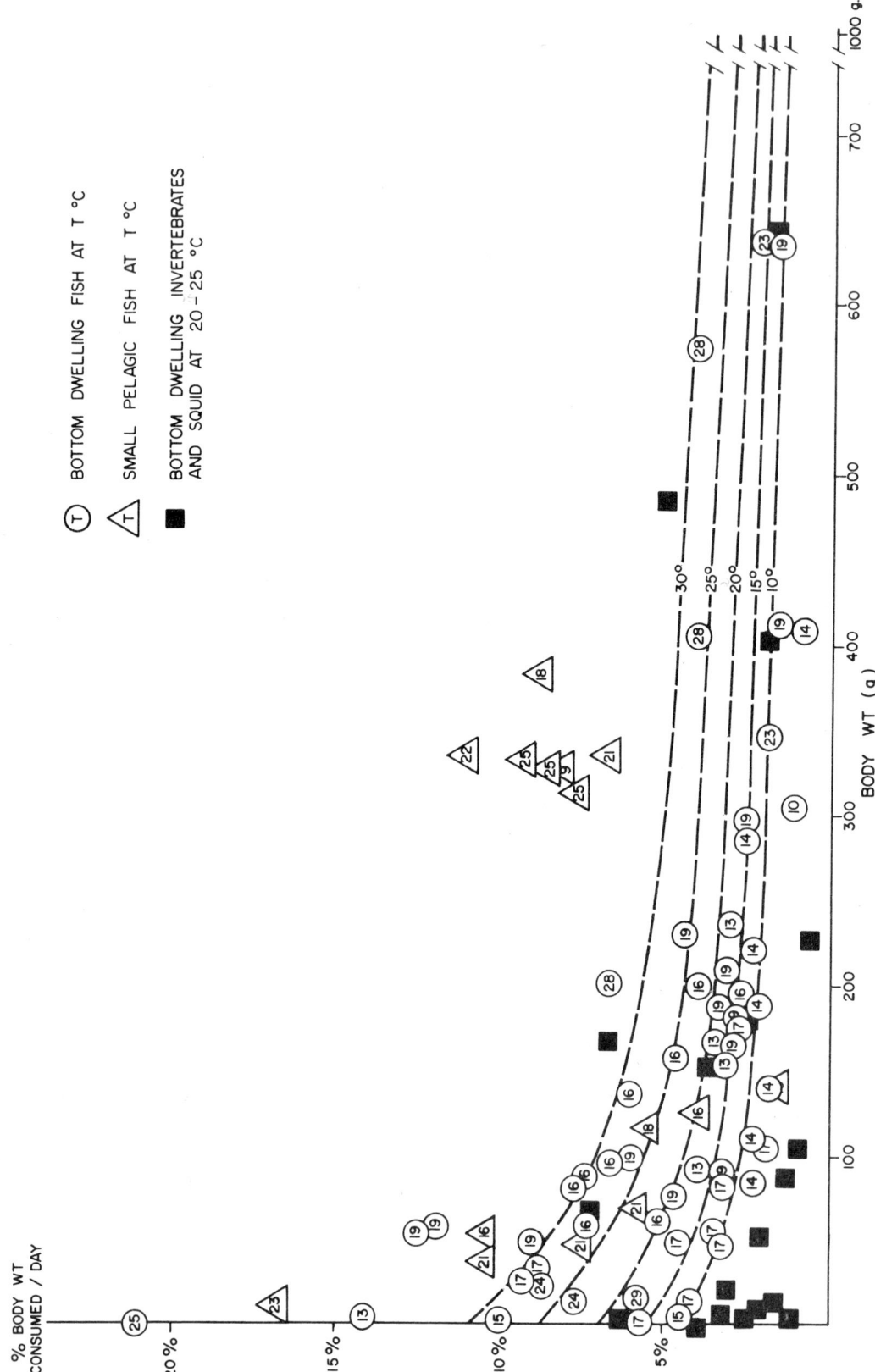

Figure 67 Relationship between body weight (g) and ambient temperature, and the consequent consumption rate (% body weight per day) for a range of fish and invertebrate species (data from Conover, 1978). The multiple regression shown by dashed lines is fitted to the demersal fish data only. (Points for pelagic fish and invertebrates added for comparison only)

In relation to the data set used, although it is not always specifically stated, we can probably assume that the majority of the data points are derived from tank experiments (especially since the size range is limited), and that therefore food supply was not the limiting factor, but activity probably was. In nature, food is more liable to be limited and/or patchy in availability, but we may not be greatly out of line to use values obtained from equation A to obtain first estimates of natural feeding rates, recognizing that these are possibly lower bounds and that activity levels will determine the more realistic values.

Olson (personal communication) provided the following table (Table 17) of daily ration estimates from field studies which shows the importance of seasonality in some species. The dominant role of body size in defining factors of ecological importance can also be illustrated by Figures 11 and 12, which shows the relationships between intrinsic rate of increase in population size (indirectly a function of feeding rate) and body size, and the relationship between body size and the theoretical concentration of "particles" (including living organisms) in sea water.

The relationship suggested by Table 17, although based on a rather limited data set, suggests that over the usual commercial size ranges, a suprisingly limited range of feeding rates of from 1-5 percent body weight per day should cover most of the variation for most (relatively inactive) demersal fish species (Figure 67). Consideration of pelagic fish species is limited by the available data set, but for two species for which we data are available (a scombrid and a carangid respectively) we have:

Pneumatophorus japonicus feeding on Engraulis japonicus	Seriola quinqueradiata feeding on Trachurus meat
$\bar{W}$ = 334 g	$\bar{W}$ = 323 g
$\bar{T}$ = 21.5 C	$\bar{T}$ = 24.6
r_T = 11.1	r_T = 8.6

The feeding rate in the two cases is 4.2 and 2.7 times that predicted for less active fish of the same size at the same temperature. This suggests that the feeding rate of active small-medium sized pelagic predators (e.g., mackerel, jacks), can be 2-4 times greater than that for comparable demersal fish in the same area. This differential is even greater for the most active oceanic predators, the tunas (Sharp and Francis, 1976). These fish provide a practical upper limit to the feeding rate, and from a model based on skipjack active metabolism measurements at $24°C$, Sharp and Francis (1976) give a range of estimates of feeding rate from 11 to 21 percent of body weight/day for a 1 kg yellowfin. This is quite outside the range of our observation set, and suggests that extreme caution be used in predicting feeding rate for organisms (e.g., tuna and oceanic squid) where locomotory activity takes up a high proportion of the energy budget.

The feature that is most striking about the data set for bottom fish, is the relatively small variation in feeding rate at any given size, and the lack of trend in feeding rate as a percentage of body weight from 30 g upwards. Despite the few data points above 500 g body/weight, feeding rate per body weight of larger fish in the same environment can be expected to be lower, but not greatly so.

The implications of these observations for population dynamics and predator-prey relationships deserve some consideration. Firstly, at a given relative abundance of a predator and its prey, the following points can be made:

(1) The natural mortality M is not likely to vary rapidly with individual predator size for a given predator/prey biomass ratio, above the individual size of prey consumed (20/50 g?) by 300 - 500 + g fish. (This tends to support the conventional assumption of a constant M for exploited size ranges, often made in population dynamics).

(2) The natural mortality rate for a given organism will generally go up with temperature. This is already seen in the relationship postulated for M by Pauly (1978) from examination of a large body of data 1/ namely,

$$\log_{10} M = -0.2107 - 0.0824 \log_{10} W_\infty + 0.6757 \log_{10} K + 0.4687 \log_{10} \bar{T}.$$

1/ This equation suggests that temperature is more important than body weight in determining the values of M_1 and this is generally in accord with our results. What is less clear from this formulation is what would be the effects of changes in abundance of predator on M for its prey species

Table 17

Daily rations (percent of body weight) of fishes calculated by several investigators using field data, ordered by increasing ration

Species	Average size (g or cm)	Water temperature (°C)	Month or season	Daily ration	Source
Northern pike Esox lucius	32-49 cm (standard length)	2	April	0.03[a/b/c/]	Diana (1979)
		14-18	June	1.8[a/c/]	
			June	3.1[c/]	
American plaice (age 11 females) Hippoglossoides platessoides	256 g	2-3	Mar-Apr	0.08[c/]	MacKinnon (1973a, 1973b)
			August	1.4[c/]	
Yellowfin tuna Thunnus albacares	2 629 g	23-32	All	0.61	Olson (present study)
	8 603 g			0.92	
	22 655 g			0.76	
Walleye Stizostedion vitreum vitreum	219 g	17	June	1.0	Swenson and Smith (1973)
	354 g		September	3.0	
Several centrarchids	40 cm-pike 16 cm-others	6-26	Summer	1-2	Seaburg and Moyle (1964)
Mako shark Isurus oxyrinchus	63 000 g	18.8	–	3.2	Stillwell and Kohler (1982)
Diamond turbot Hypsopsetta gluttulata	96 g	12.5-24.0	All	3.8	Lane (1975) Lane et al. (1979)
Perch Perca fluviatilis	>20 cm (total length)	–	June-July 1971	6.5	Thorpe (1977)
			August 1971	5.2	
			September 1971	3.2	
			June 1972	4.7	
			August 1972	4.3	

a/ Males

b/ Females

c/ Given in kcal/kg/day or kcal/fish/day – converted to % body weight/day assuming 1 000 kcal/kg

3. The natural mortality rate exerted by a stock of pelagic predators on a given prey is likely to exceed that exerted by a similar sized population of bottom fish occupying the same feeding niche.

4. Following from 3, the proportion of a given quantity of lower trophic level productivity (e.g. of a "forage" fish) recoverable from a stock of demersal predators is likely to be higher than by harvesting a pelagic stock sharing the same "forage" species.

Environment and recruitment

Most juvenile and adult fishes have in general more options available to them than do the less mobile early life history stages. A set of questions that constantly reappear are those about relationships between temperature, larval fish development time, and predation.

The absolute temperature of the environment within the normal range that a species is adapted to, is not critical to survival of larvae in most cases, since although for metabolic reasons, intermediate to cool temperatures within a larval fish's normal temperature range promote better survival for individual larvae given adequate food, predators in warmer habitats also follow the same thermal laws. Hence invoking Q_{10} as a biological quotient which can be applied in many situations, means that the logarithm of a predator's activity increases in proportion to increased temperature, so that predators are especially voracious in a "warm" habitat; cancelling nearly all benefits a larval fish might derive from developing rapidly from egg to larval stages.

In conclusion, upon hatching fish larvae are faced with a trade off between higher rates of development at higher temperatures, and higher rates of predation for precisely the same reason. The conclusion seems to be developing, particularly in the tropics, that lower to average temperatures may even be an advantage for development of fish larvae. The following table gives the essence of the problem:

Table 18

Relationship between temperature, and the activity and respiration/feeding rate

Habitat Temperature (^{o}C)	Activity Index	Respiration/Feed Index
31	1.414	283%
30	1.320	230%
29	1.231	187%
28	1.149	152%
27	1.072	123%
26	1.000	100%
25	0.933	81%
24	0.870	66%
23	0.813	65%
22	0.758	43%
21	0.707	35%

If we take the 26°C isotherm as a likely centre point for egg/larval hatching and development in the tropics, and use the requirements in food/respiration at this temperature as a baseline, the effects of temperature alone on activity level can be indexed as a simple function of Q_{10} and are shown to range from + 41.4 percent above the 26°C value with a five degree Celsius ambient temperature increase, or conversely to be only 29.3 percent with a five degree decline below 26°C. However, the energy expenditure based on activity alone for a continuously swimming larval or post-larval fish varies as the cube of the swimming speed. As one can quickly see, this results in dramatic food/respiration differences.

18. ESTIMATES OF NATURAL MORTALITY RATES FROM WHOLE SYSTEM VARIABLES

In conventional fisheries analysis of the dynamics of single species, we are accustomed to assuming that natural mortality rate M can be treated as a constant through time for the exploited phase of the life history. This assumption is necessary in order to solve the catch equation, even though it can usually be shown by sensitivity analysis that the yield predicted by this equation is particularly influenced by variations in this difficult to measure and varying parameter. The assumption of a constant M is tolerable for single species assessments, and may be seen as a logical consequence of the slow rate of change of feeding rate with size for predators above a size of 50 g or so, shown in Figure 67. This conclusion is valid however if relative abundances of predator and prey remain roughly constant, which is not always the case. From the point of view of an attempt to understand multi-species interractions, it may be a logical alternative to the 'static' assumptions of (e.g.) the ECOPATH model to regard the value of M as an output variable resulting from the relative abundances and feeding rates of predator and prey species in the ecosystem. The main point to be borne in mind is that concluded at the recent ICLARM/CSIRO workshop on Theory and Management of Tropical Multispecies Stocks, namely that for fisheries where multi-species effects are beginning to seriously bias single species assessments, we need some tools, however crude, that allow us to base our conclusions on events at the whole system level. An approach to modelling M as an output parameter, determined by the intensity of predation, (in its simplest terms, a function of relative predator and prey abundance and predator feeding rate), would be a useful preliminary to understanding system dynamics.

Another option that may be used to simplify multispecies models, is to express growth, mortality and migration rates in terms of biomass as opposed to numbers (e.g., Laevastu and Larkins, 1981). These latter authors noted that in data-poor areas, the use of <u>ponderal</u> rates can be preferable to number-based models, even if this necessarily sacrifices an ability to integrate results easily with age-structured models. These authors propose definitions for e.g.:

Biomass (B_n) in year n:

$$B_n = B_{n-1} \exp(g_n - Z_n)$$

where: Z_n is the mortality rate in year n,

B_{n-1} is the biomass the previous year, and

g_n is the exponential rate of biomass growth in year n.

Similarly, for natural mortality caused by predation:

$$M_p = -\log_e (1 - D/\bar{B})$$

where D is biomass consumed by predation and $\bar{B}$ is the mean biomass in the year. These concepts are developed a little further in the next section in suggesting simple indices, particularly for ranking the likely impact of different predators on the natural mortality of the prey.

The relationship between predator and prey biomasses and the natural mortality rate of prey

The relationship between predation and natural mortality rate in the prey population was expressed as a function of predator biomass by Munro (1979, 1982) as:

$$M = M_x + gP \quad \dots\dots\dots\dots\dots\dots\dots (1)$$

where P = biomass of predators, and M_x is mortality due to natural causes other than predation (e.g., disease, senescence, etc., where this <u>does not</u> result in prey consumption by the usual predators). Evidently in this context, M_x is likely to be very small, since senescence and disease usually result in an increasing probability of predation, and from a trophic viewpoint these two types of factors simply act to facilitate transfer of biomass from prey to predator, and we would probably be justified in writing:

$$M \leq gP \quad \dots\dots\dots\dots\dots\dots\dots (2)$$

This second equation assumes however that predator biomass P act in an analogous way to fishing effort (f) in the equation $F = qf$. This is somewhat misleading, since equation (2) either assumes a constant abundance of prey, or that a given biomass of predators acts "catalytically" to induce a given prey mortality, independent of the biomass of prey. In view of the trophic needs of the predator, prey biomass must also be a term in the equation; especially since from the predator-prey theory, we know that predator + prey biomass each tend to oscillate out of phase with the other. Evidently, as prey biomass declines with above average predator biomass, M must increase if the predator's food requirements are to be satisfied. However, a rather intuitive yet admittedly crude approach can be formulated, again looking in the simplest case at an obligate relationship between a single prey and its single predator in an unexploited system, and ignoring problems of availability. One can postulate after Laevastu and Larkins (1981), a relationship between predator and prey mortalities such that the proportion of prey consumed in an interval will be a function of the relative predator and prey biomasses or densities within any given framework.

Assume first that the food requirements of the predator in unit time can be expressed as a simple fraction (or multiple) of their own biomass = aB_{i+1}.

At the end of the time interval, ignoring the effects of growth of predators and prey, the original prey biomass B'_i is reduced to $(B'_i - aB_{i+1})$, and the corresponding rate of decline in biomass due to predation M'_i is then:

$$B'_i - aB_{i+1} = B'_i e^{-M'_i}$$

or
$$M'_i = -\log_e \left[1 - \frac{aB_{i+1}}{B'_i} \right] \quad \ldots\ldots\ldots\ldots (3)$$

This formulation of course assumes that we can define mortality rates in terms of biomass rather than numbers. Such a rate (M') has been referred to as a ponderal rate, and is not necessarily identical with the numerical natural mortality rate, M. This approach may be acceptable as a first approach, but size-specific predation and age composition of predator + prey will need to be allowed for, as suggested earlier. Also the assumption of a constant feeding rate implied by constant $\underline{a}$ ignores density-dependent changes in energetics of feeding with season and size, and takes no account of the different energies needed to capture different prey species, and the different calorific values of different prey. Despite this, equation (3) appears to be a reasonable point of departure for a very simple treatment of the problem, and as we have noted, a similar expression has been used by Laevastu and co-workers.

Looking briefly at some of the difficulties with the simple formulation outlined above, one may first suggest that as long as we discuss predator-prey relationships in terms of quasi long-term average abundances of species in the ecosystem (the "equilibrium assumption" that underlies most currently used fisheries theory), implying that predator-prey ratios remain relatively stable, then the surplus production due to growth and reproduction of the prey not harvested by man, will on average, be roughly equivalent to the increment in biomass of predators, (after allowing for conversion efficiency of prey to predator biomass). It may be acceptable therefore under these assumptions to substitute mean biomass in equation (3). This formulation:

$$M'_i = \log_e \left[1 - \frac{a\bar{B}}{\bar{B}'} \right] \quad \ldots\ldots\ldots\ldots (4)$$

then gives us a simple first-hand way of deriving orders of magnitude for M, given some idea of trophic interrelationships (i.e., a food web), and an idea of the relative biomass of predator and prey; such as, for example, might be obtained from stratified trawl or acoustic surveys if these can be shown to be measures of relative biomasses.

Prey mortalities due to predation

Looking at interelationships between a single prey and its single predator, and ignoring problems of availability, one can postulate relationships between a predator and the mortality it exerts on its prey, such that the proportion of prey consumed in an interval will be a function of relative abundance of predators to prey biomass. On the other hand, if prey abundance falls below predator food requirements, prey switching, migration of predators from the system, or even in the extreme case, natural deaths of predators due to starvation could occur, which cannot be accounted for in this simplistic approach.

In all of these circumstances, the above simple mathematical approach breaks down. However, assuming that Figure 67 in this paper gives some idea of the magnitude of the mean annual feeding rate (i.e., a' = 365 a), then it is instructive to see what order of magnitude or M is predicted from equation 4.

Consider a predator-prey pair where a predator A with mean individual weight of 700 g and mean biomass of 1 000 t is the only predator, and feeds exclusively on species B', a prey species with biomass 15 000 t, at ambient temperature $10^\circ C$. Applying equation 4, the predicted annual value of M for species B' due to predation would be:

$$M_{B'} = -\log_e \left[1 - \frac{(365 \times 0.0156 \times 1\,000)}{15\,000} \right] = 0.48$$

Assuming the ambient temperature increased to the level which doubled the feeding rate of A:

then $\quad M_{B'} = 1.42$

and if the biomass of B' is now reduced to 600 t by fishing (prey biomass and temperature constant):

$$M_{B'} = 0.61$$

Obviously the actual values resulting must be considered with some scepticism, but this approach may be useful in predicting effects of important changes in relative biomass or temperature, or in comparing impacts of resident and migrant species on a common prey. For this latter purpose, the equation could presumably be modified to give some idea of the impact of seasonal predation by multiple predators on the value of M for prey species B'. Thus, for i = 1, 2, 3 ...n predators of biomass B_i, each of which spends a fraction Δt_i of the year feeding on common prey species B' at a daily rate a_i, and where P_i is the proportion of the diet made up by B', then if the same logic holds,

$$M_{B'} = -\log_e \left[1 - \frac{365 \sum_{i=1}^{n} a_i \Delta t_i P_i B_i}{B'_B} \right]$$

Thus, if a biomass of 10 000 t of prey species A, and three predators C, D, and E make up part of a simple "assemblage production unit", with characteristics as below:

	Daily Feeding Rate	Proportion of year feeding on A	Proportion of A in stomach contents	Biomass (t)
C	0.02	0.3	0.20	100
D	0.03	0.3	0.10	500
E	0.06	0.8	0.15	850

What is the predicted value of M? From the above $M_{B'} = 0.28$.

Perhaps in conclusion, the value of this type of approach is more that it allows a prediction of the effects of certain types of population change on the unknown or known values of M, rather than as a method of estimating the value of M itself. The approach also seems to have some merit in providing first estimates of parameters to simple models such as the ECOPATH approach, and in comparing relative impacts of residents and transients for example, on a common prey. It also, seems to provide an approach to tackling the 'non-equilibrium' problems in fisheries, since a change in natural mortality rate of species in a food web is one likely cause of the changes in species composition often seen is some heavily fished systems.

19. SOME IDEAS FOR REPRESENTING FOOD WEBS IN THE FISHERIES CONTEXT

Although diagrams can only supplement numerical models, one of the aspects we have emphasized in this report is the utility of diagrammatic approaches for communicating information on ecosystem interactions and the food web. Despite the variety of conventions used in depicting flow diagrams however, there are two aspects of trophic relationships in the sea that are not normally represented, although they are of major importance. The first of these illustrates the spatial component of energy transmission - a major theme of this report being the food web as a "dissipation structure", in which specific areas and habitats play a key role in biomass production and dissemination (see Fig. 3 for an example of the spatial representation of food web linkages, and Laevastu and Larkins (1981) for a discussion of the importance of spatial distributions and hydrographic factors). A second characteristic of most marine food webs that is rarely represented, is the fact that most components at all trophic levels start off life at close to the same size range, and apical predators for example, almost inevitably move upwards trophically in the course of their life history. We have concluded that size relationships (as noted for spatial considerations above), are essential to understanding ecosystem processes, but are rarely represented in food webs. All of this prompts us to tentatively suggest an approach to food web representation that takes these spatial and size-related elements into account, together with that other important information element; the species biomass.

Figure 68 attempts to portray the relationship between dominant (A) and subdominant predators (B) and their common prey (C) in a manner that makes explicit the size interrelationships as well as the biomasses involved (both shown on a logarithmic scale). The possibility of "reciprocal" predation by adult prey on the juveniles of their predators (the cod-herring example), is easily accommodated by this convention. It is also possible to sum all biomasses (not their logarithms!) laterally to show the biomass spectrum of the subsystem in question. The "kite" diagrams in Figure 68 for each species obviously are 'doubles' of the biomass spectrum for each species. Such biomass spectra, (here, obviously incomplete for lower trophic levels), could be portrayed prior to and after exploitation of one or more components, and are analogous to the pyramid of numbers shown in Fig. 2.

20. SINGLE SPECIES MANAGEMENT IN AN ECOSYSTEM CONTEXT

Differences in Vulnerability to Fishing

Much of fisheries theory as we have already noted, still operates on the assumption that individual components can be managed separately from their preys, predators and competitors, and the persistence of this school of thought and practice for over a half century since Baranov's earlier work, shows that this may not always be a totally incorrect perspective in some ecosystems for much of the time. It may be worth speculating that this could be true because of the 'damped' nature of many marine systems, where individual food web components tend to overlap with others in the same system giving a certain resilience to the food web, in the face of the exploitation of one component; or perhaps because at moderate levels of exploitation, production by the exploited species is in fact increased by fishing.

Critical Habitats and Refugia

In trying to put an order of priority to our concerns with species management in an ecosystem context, the complexity of marine ecosystems makes it seem problematic that except for very simple systems, we will be able in the near future to guarantee a given impact from a particular intervention at a specified 'node' in a food web. Thus, although for example, by selective fishing, we can in theory, remove "undesirable" components, this would not only be difficult to justify economically, but from the few examples where this has been tried, such as the attempt to remove dogfish (a low-value predator on salmonids) in the northeast Pacific by a 'bounty' system, or to control starfish and gastropod predators in shellfish culture, these efforts are rarely crowned by lasting success, even if (as often happens), a market for these "pests" subsequently develops.

Large-scale experiments have in fact been carried out in ecosystem manipulation all over the globe, namely by single species commercial fisheries. Although these have not been aimed at species eradication, in a relatively low proportion of cases globally have measures other than a decrease in economic yield at low abundance, protected the target species from possible extinction. There have of course been cases where valuable species have been driven to extinction or rarity (especially those with high vulnerability and long-life spans), but there have been a surprising number also which have shown a significant degree of resilience to the effects of such interventions.

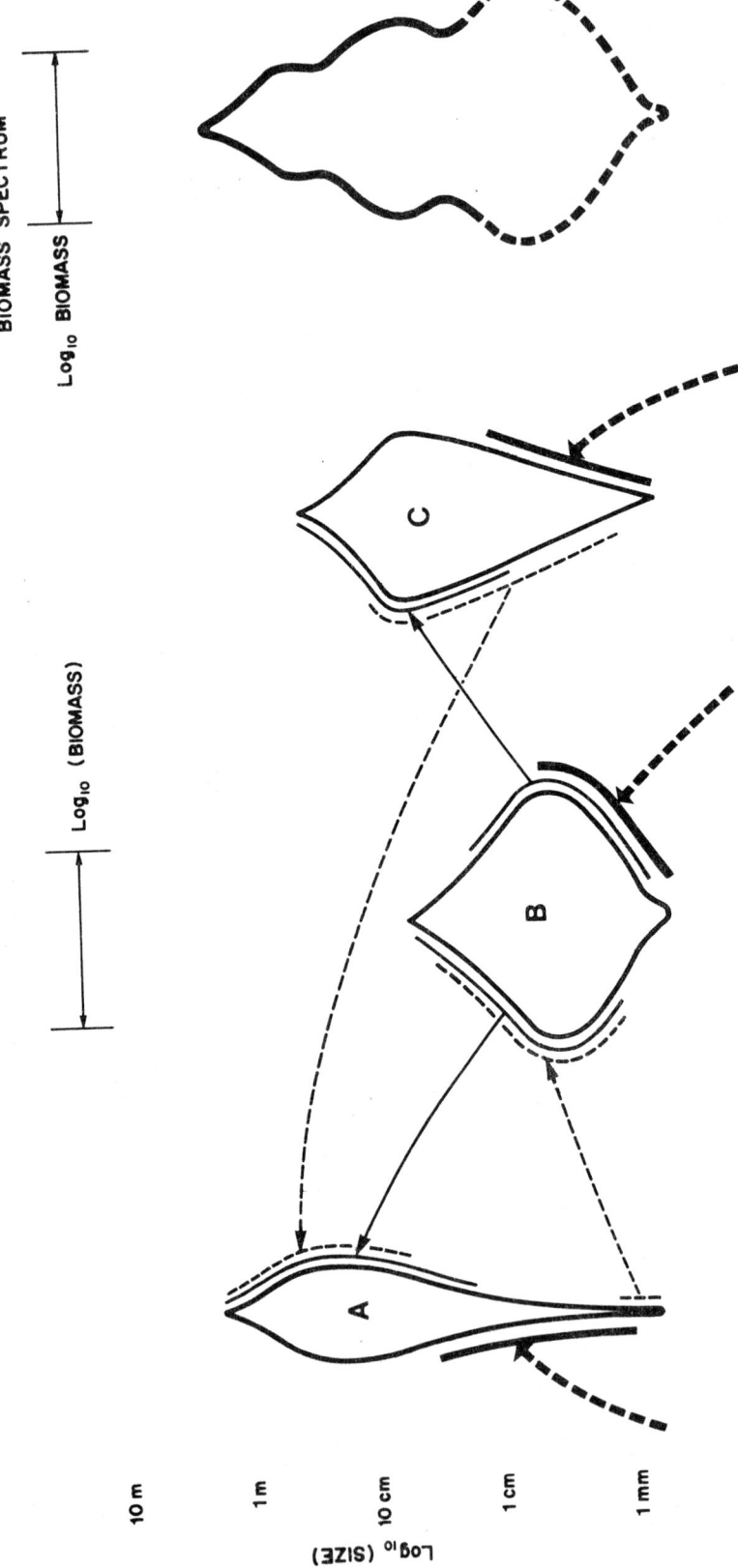

Figure 68 Showing a food web for three species components, with logarithms of biomass shown as "width", and arrows indicating trophic exchanges between given size ranges. Reciprocal control of larval recruitment of "predator" A by (planktivore?) "prey" B is shown; also the biomass spectrum for the whole subsystem. (Arrows without origins indicate unspecified units lower in the food web)

More detailed examination of both categories seems to show that two types of intervention are more likely to lend to drastic stock declines and overfishing:

(a) The exploited phase of a population goes through a period (usually when the stock is concentrated in a given area), where it is highly vulnerable to fishing. Such areas may be called "critical habitats". Commonly cited examples of critical habitats are the spawning grounds of species such as cuttlefish, herring, grouper and marine turtles where much of the adult biomass is concentrated in a small area over a limited period of time, and hence vulnerable to high levels of exploitation where fishing success is largely independent of stock biomass.

(b) The species passes through a phase in its life history associated with a particular habitat, e.g., mangrove swamps and shallow lagoons for some shrimp and crab species, grass beds or macrophyte stands for some lobster species, etc. The "health" of the "substrate" species then has an overriding influence on the associated organisms.

In the contrary sense, where a species resists heavy exploitation, it may be because capture efficiency is proportional to density, or because a "refugium" exists which protects the species from heavy exploitation over part of its size range and geographical extent.

A species "refugium" is less easy to define, but may consist of an area of rocky untrawlable bottom, or for a small boat fishery, an area of strong currents distant from shore. It may also occur as a result of fishing with gear such as traps or gillnets which have a particular size selectivity, such that the dominant mesh size or entry hole dimension of the gear forms a "refugium" for those animals that avoid capture long enough to grow to larger, mature, sizes.

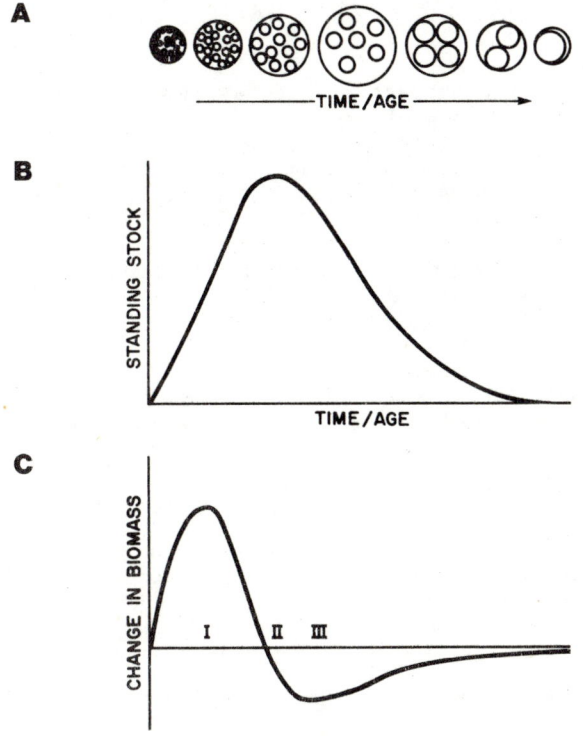

Figure 69   (A) Showing evolution of a cohort with age from many small individuals to a few large ones

(B) Standing stock of the biomass of this same cohort with time or with age

(C) The rate of change in biomass (dB/dt) with age (II marks the age at maximum biomass, given the growth and mortality rates that apply, and I and III mark respectively, the times at which the biomass is growing and decaying fastest

(Redrawn from Alverson and Carney, 1975)

Biomass at Size Distribution and Stock Recruitment

The importance of information on the length spectrum of a species for analysis of population trends, and estimating population parameters, has become more appreciated in recent years, and the equivalent in terms of biomass is of corresponding importance for trophic calculations. The biomass distribution with size and age for a species shows an increase to a maximum at the age/size when growth in weight is just balanced by mortality due to all causes, before subsequently declining with size (Figure 69). For exploited resources, this point of maximum biomass would be the ideal size for instantaneous harvesting. In practice, however (and here the yield per recruit methodologies of e.g. Beverton and Holt, 1957 came into play), it is usual for a given natural mortality rate (M) assumed constant for exploited sizes), to calculate the rate of exploitation that would provide the maximum yield per recruit for a given size at first capture, or the impacts of a change in mesh size or size at first capture. The possible impact that heavy exploitation might have on prey or predator populations of the species in question, is not considered in the approach.

One of the other limitations of this type of calculation which has major practical implications, is that the predicted effects of changes in size at first capture do not always take into account whether or not a species begins to be exploited before sexual maturity: this could of course have an impact on the average level of recruitment, if in the extreme case, exploitation is heavy and begins before maturity. This type of effect is dealt with by so-called stock recruitment models (see Ricker 1975 for details). These models which usually attempt to predict the average level of recruitment from a given spawning stock size alone, don't usually take into account variations in the physical and biotic environment the spawners and pre-recruits are subject to, and as a consequence show a high variance in predicted number of recruits. What is important to note from the perspective of trophic analysis however, is that the 'loss' of commercial-sized animals in yield per recruit evaluations due to natural mortality (the proportion M/Z of all deaths) returns material to the food web. This is true also for spawner-recruit models: thus although the number of recruits successfully produced per unit biomass of spawners drops off with total spawner biomass the 'losses' in gametes and pre-recruits implied here, re-enter the food web; but at a generally smaller prerecruit size.

In practical terms therefore, although we should not abandon the useful single-species methodologies that have been developed over the last 50 years or so, some attempt should be made to place these types of analyses in their correct ecological context, i.e., every single species assessment should ideally be preceded by a qualitative statement about:

(a) The physical system, and what we know of its stability, seasonality, extent and productivity.

(b) What we know about the other major biological components in the system and their interactions (trophic or otherwise) with the species in question.

(c) The variability in production of the resource in question and of the other key resources considered under (b), and some idea whether the present biomass is currently in the upper, middle or lower quartiles of the historical range.

Following the single species assessment therefore, should come a qualitative statement which puts the 'ifs and buts' around the numerical conclusions, so that the manager can make some judgement as to how applicable it may be, and for how long. This statement could draw upon a body of experience from outside the particular fishery in question, and one may suggest that it would not be premature to look retrospectively at the history of industrial fisheries over the last few decades: a cursory examination (e.g., Caddy and Gulland, 1983), may suggest that the "Theory of Catastrophes" rather than the concept of "population equilibrium" is a more applicable one in some cases!

21. IMPACTS AT THE ECOSYSTEM LEVEL

Several of the earlier sections have dealt with ecosystem components and their characteristics, which may have led to the mistaken impression that the properties of an ecosystem are simply the sum of the properties of their individual components. This is not true, since there are "emergent properties" of ecosystems which cannot be forseen from those of the individual components. Thus, on the most practical level, the multispecies yield from a system is significantly less than the sum of the maximum yields of the individual species if harvested alone. Mann (1982) notes that static diagrams of food webs, particularly compartment models, are doomed to failure as quantitative predictors, largely because the emergent properties of such systems are not revealed, although these diagrams still remain useful descriptors of how food and energy flows through the system.

This leads to consideration of the main alternative approach; namely the application of empirical approaches to marine ecosystem assessment. We could look for example, for a marine equivalent of the morphoedaphic index (Figure 70); which estimates freshwater fish production in lakes as a function of easily measured physical factors. This of course has two major problems: the first being that marine systems are all 'open ended' to varying degrees; the second, that changes in the 'state' of the subsystem concerned on which the index is based, throw the time series of catch and environmental information out of gear forcing a re-examination of underlying structural questions.

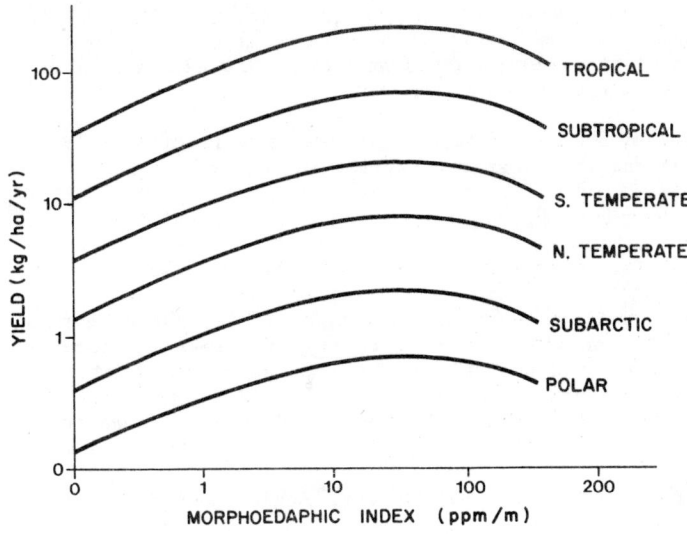

Figure 70 Predicted yield curves as functions of the morphoedaphic index for sets of natural lakes in six of the world's climatic zones (Adapted from Henderson, Ryder and Kudhongania, 1973)

A variety of ecologists have addressed emergent properties of ecosystems, and although perhaps the most fundamental law of ecology is that the more sweeping the generalization, the more exceptions to it will be found, the following perspectives are offered which will hopefully prove useful in discussing responses at the ecosystem level.

One place to start is with the statement of Margalef (1978) who suggested that "all or most of the ways in which man interferes with the rest of nature produce coincident or parallel effects"... ....."diversity is reduced and the ratio of production/biomass is increased". If this is true, it certainly makes for a simplification of many apparently diverse problems, and the idea has been developed further in what appears to be a new field, namely "stress ecology" (Barrett, Van Dyne and Odum, 1976). Some of the basic concepts and terminologies used here (Rapport, Regier and Hutchinson, 1985) derive from work on the response of individual organisms to stress; which although not a precise term in this context, embraces the idea of a perturbation imposed on the system from the "nominal" state; presumably taking it away from an original "equilibrium" or "pseudo-equilibrium" state of the system, in the sense in which these terms are used in population dynamics. Rapport, Regier and Hutchinson (1985) point to the three stages postulated by Selye (1974) for the response of an individual organism, namely "an alarm reaction", then a stage of "resistence to stress" and finally "exhaustion" or passive compliance with the stress, as having some parallels from a cybernetic point of view, at the ecosystem level (Figure 71).

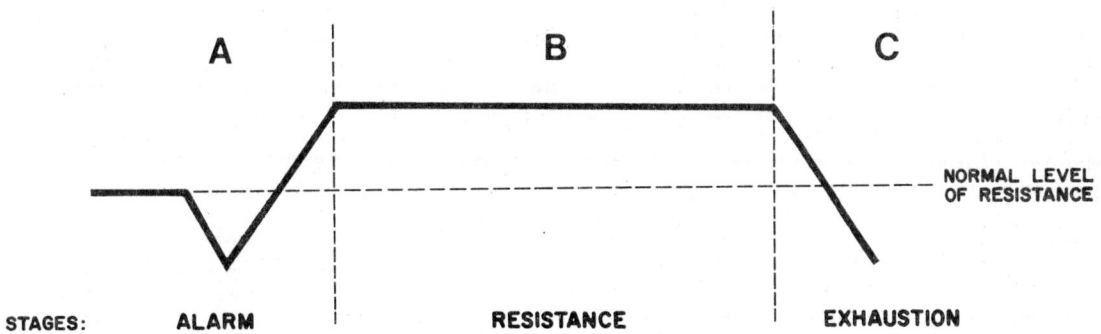

Figure 71 Three phases of the general-adaptation syndrome in organisms: (A) alarm reaction; (B) stage of resistance; (C) stage of exhaustion (From Rapport, Regier and Hutchinson, 1985)

Five types of stresses seem to include most human impacts on ecological systems, which are not exclusive of human interference, since natural causes may produce similar effects:

(1) Harvesting

(2) Pollutant discharges, (perhaps distinguishing man-made toxins where this generalization does not apply, from pollution by natural organics)

(3) Changes of terrain, bathymetry, degree of shelter etc., (e.g., the effects of shoreline development, land reclamation, dredging and marine gravel extraction, etc.)

(4) Introduction of exotic species

(5) Devastating events (hurricanes, oil spills, coastal pollution, etc; (but see (2))

Under (5) the above authors include wars as well as natural catastrophes: (it may be noted here in passing however, that our best information on the rate of _recovery_ of fisheries systems from stress, which has led to much current population dynamics theory, came as a result of the virtual cessation of fishing in the N.E. Atlantic during two world wars). The importance of degradation of the marine environment (pollution) by human activities is reviewed, e.g., in Anderson, 1981.

The diagnostic changes for ecosystem stress recognized by Rapport, Regier and Hutchinson (1985) included:

(1) an increase in the rate of loss of nutrients from the system (i.e., partial break down of the dissipation structure or food web),

(2) changes in primary productivity (as noted in upwelling zones, or when increased leaching of sediments and fertilizers takes place into coastal zones from agricultural development),

(3) increase in production/biomass ratios,

(4) decreases in species diversity,

(5) retrogression. This is defined as a shift in species composition to those organisms best adapted to new and more difficult environmental conditions; in our context, to short-lived r-selected organisms, as noted earlier. (If regarded from the point of view of ecological succession and the concept of climax in biological communities, this would imply a movement to "less mature" communities),

(6) changes in size composition, often towards smaller mean sizes and shorter life spans: again, consistent with the shorter life span of r-selected species, and the higher P/B ratio mentioned in (3). (In the contrary sense however, the failure of recruitment may lead to an _increase_ in mean size becoming an index of population stress),

(7) other signs of distress, such as the "biological distress syndrome" may occur, and could include all of the above, plus disease, changes in growth rates, and mass mortalities.

In fisheries, changes in the ratio searching/fishing time (e.g., Condrey, 1984), or in the number and location of spawning areas, may all be indicators of stressed conditions or changes in population size, distribution and abundance. The disappearance or progressive rarity of indicator species for unstressed systems, and their replacement by parallel more resistant species, may be another type of "alarm reaction" as mentioned above. These changes may also show up on the local fish market as increases in the relative price of popular (stressed) species and their replacement by alternative (formerly low grade species); again suggestive of ecosystem changes, although here changes in trade or consumption patterns may be the causative agents of higher exploitation.

Perhaps the least optimistic but quite realistic view of the potential use of current approaches to modelling trophic interactions, is that expressed by Gulland and Garcia (1984), who note in relation to the question of interaction and stability in multispecies fisheries, that although these matters deserve increased practical and theoretical attention, the theoretical studies have done little so far except "open the door on a complex image of interrelated problems". A similar view was expressed by Sainsbury (1982), who after a comprehensive review of mathematical modelling approaches up to and including the ICLARM/CSIRO Workshop in 1981, concluded that "close examination of ecological and fishery theory indicates that at, present there is no ecologically adequate model of community dynamics, and that no "coherent body of methods and rules" which might be applied to management of tropical multispecies fisheries is apparent in the approaches he reviewed. As we have noted, to some extent the complexity of the subject results from the scientific approach adopted by the academic world, which places a high priority on innovative as opposed to synthetic activities: given that as was pointed out by Kuhn, and well exemplified by ecological theory, science consists of a series of separate disciplines each of which vie competitively in the literature, and for funds in the "scientific market place".

Despite the pessimistic view expressed above as to of the possibilities in the near future of producing a generalized multispecies model, (which is itself an overambitious project given the wide diversity of multispecies fisheries see e.g., Pauly and Murphy (1982), Gulland and Garcia (1984) for examples); this does not imply that a concentrated approach to understanding <u>one particular</u> fishery will not meet with a degree of success. What we have tried to do in this review, is to illustrate some of the ideas and concepts that might be considered in such an approach. Quite evidently however, multispecies systems cannot be tackled with any degree of success unless a significant commitment is made; ideally on a broad cooperative basis.

Thus, although there are no simple recipes for success, there seem to be some general guidelines for viewing a multispecies fisheries system in an objective way that may prove useful in a given situation, even though a fixed plan of action that applies in all circumstances is not possible.

(1) The first suggestion we have is that an attempt be made to categorize the main assemblages or communities present, (separating resident species from transients), and map them on a chart of the fishing grounds. Even if this is done on a very preliminary basis, it illustrates the spatial relationships of these assemblages to the main fishing ports, and can be the basis for a fisheries statistics system that attempts to the extent possible, to determine the removals from each fishing area separately. Satellite imaging now provides a rapid approach to characterizing coastal zones, water masses and local areas of high production, and can help in this process.

(2) Where possible, the historical data base for the fishery, including ancedotal information from fishermen as well as from searches of the available literature, should help establish files for the key species, their habits, trophic interrelationships and life cycles. The occurrence of major changes in abundance and species dominance in the past, (even if qualitative), should be documented; together with area-specific information on fishing effort, fleet size, fishing grounds and past removals, and such historical data as are available on catch rate and size composition.

(3) Where possible also, without making this a major program in manpower-limited situations, limited observations on stomach contents of key commercial species of known sizes, seasons and areas, may be collected in order to build up some idea of species interactions in the system.

(4) One key aspect of the application of ecological concerns to fisheries problems, is the identification of the major sources of biological production, and the areas where these centres are located in relation to human activities.

(5) One tentative suggestion that arises in this regard, is that an attempt be made to map critical habitats and the centres for ecosystem "dissipation structures", in the manner portrayed on the cover of this report, with a view to assigning a high priority to their conservation, even if the details of food web interactions are not yet resolved. This type of action may be regarded as preserving the 'options for the future', when the necessary fisheries ecological studies have been carried out.

<u>Strategies of harvesting multispecies ecosystems</u>

The estimates of total fish production based on commercial yield statistics, are probably underestimates of the potential yield which would result if the fishery were exploited with the balanced diversity of gears needed to evoke the highest yield (Marten and Polovina, 1982). What strategy is implied by this statement?

In theory, a food web could be maintained 'in balance' by fishing each component in proportion to the rate of natural predation it is subjected to.

This 'utopian' strategy for exploiting a food web might be considered as equivalent to satisfying the requirement:

$$\frac{Y_1}{M_1 B_1} = \frac{Y_2}{M_2 B_2} = \frac{Y_3}{M_3 B_3} \ldots = \frac{Y_n}{M_n B_n}$$

i.e., a constant proportion of the total production $p_i = M_i B_i$ for species $i = 1, 2, 3 \ldots n$ is removed as yield (Y_i), where M_i is the natural mortality rate, and B_i the biomass. This can be shown to be equivalent to setting the ratio F_i/M_i at some constant level for each species, in what is effectively a form of effort control by species.

As for any simplistic scheme, a number of serious objections can be raised as to its practicality:

How could you measure all the parameters accurately, let alone control the species-specific rates of fishing?

What would be the 'safe' levels of F_i/M_i to use?

Caddy and Csirke (1983) show that F_{MSY}/M values in the literature vary widely, and that larger values well below 1, and even as low as 0.33 for some species groups, would seem appropriate from the limited parameter sets available.

It is obviously dangerous as we have seen, to assume that the abundance, or even nature of the components of a food web are constant: with too high an F/M ratio, species replacement would seem a distinct possibility, especially since when mean sizes of food web components change, this changes the effective position of a species in the food web, and makes surplus food organisms available for colonization of the ecosystem by a 'new' predator or species group.

Since many larger species of marine fish pass through sequential trophic levels in their life history, the idea that we can increases the total production of the system by removing the top predators, and then harvesting the 'forage' species lower down in the food web more intensively for greater yields, does not seem to work in practice in most cases.

It appears (e.g., Regier and Henderson, 1973, Ursin, 1982) that as fishing increases on a dominant or apex predator, its role as a controlling agent is replaced, at least in part, by the next predator down; the older individuals of which are able to move into a higher trophic category than formerly. Certainly, experience in large lakes (Regier and Loftus, 1972) shows that cropping the large predators may increase yield by less than 100%, as opposed to the 5-10 fold increases predicted by laboratory experiments assuming a fixed fish dietary composition.

An alternative strategy of fishing all species indiscriminately by broad-spectrum fishery gear such as trawls, should favour broad niche, opportunistic species with high fecundity. It seems clear from experience however, that the response of the system to fishing is less easy to predict, given that fishing as an economic activity may depend for its success on which of two "broad-spectrum" predators is favoured: the lower valued sculpins or the higher valued cod-like fishes?

Pauly (1979) summarized some of these "food web management options" in a semi-humorous form, among which the following frequently occur in the real world, and some of which, at least implicitly, have been referred to earlier. Using Pauly's descriptive titles, a selection of these are:

(1) The "Null" strategy:

Here the resource is left in a virgin state.

(2) The "Tuna" strategy:

Basically, this is the strategy of skimming off the MSY from the peak predators by fishing methods selective for large fish. As noted earlier, by May et al. (1979), such a strategy, though not providing the maximum ecosystem yield, is sustainable.

(3) The "North Sea" strategy:

Pauly (1979) characterizes this strategy as the systematic overexploitation of peak predators until predation on prey species becomes negligible. This should lead to an increase in prey species biomass to a maximum; (see however Ursin's hypothesis!), but this is prevented by then fishing the prey species; thus transferring their production to fish catch. This may be the most productive strategy in northern latitudes where there are larger number of generalized feeders, and may even be sustainable in those areas.

(4) Gulf of Thailand strategy:

This consists of fishing both predators and prey with very small mesh gear at high effort levels. In the final analysis, this results in collapse of both predator and prey stocks, and increases in plankton and benthic biomass, and of certain generalized, rapidly-reproducing r-selected species, such as squid and small flatfish.

Other variants may be considered, but these last three strategies in some ways span most of the feasible spectrum, with the addition of the "Null strategy" which strangely enough is practical in association with one of the others, namely, the preservation of part of the virgin ecosystem intact by means of Marine Parks or Reserves, particularly for critical habitats.

In a previous section, we have postulated one further strategy namely the "Utopian" strategy, which would seek to balance overall exploitation, by selective fishing of all food web components so that the multispecies balance is preserved. Perhaps the Japanese inshore fishery is closest to a real-world example of this, where a tradition exists of using a much wider spectrum of marine products then in some other parts of the world. This may prove a useful concept to keep in mind when discussing alternatives for multispecies management.

In the real world, a mixture of strategies may be the best approach. For example, a combination of (1) and (3) or or (1) and (4) might be considered, where some critical habitats are entirely closed to fishing, or at least closed on a seasonal basis.

In attempting to compare the feasibility of ecosystem manipulation (or the vulnerability of ecosystems to change), there is a strong temptation for open sea ecosystems to place the various types of interventions we have discussed in the following order, by feasibility and impact:

Feasibility	Type of manipulation	Vulnerability
High	- Impacts that affect critical habitats	High
"	- Impacts on 'substrate' species	"
Medium	- Impacts that affect major sources of biological production (dissipation structures)	Medium
Low-High	- Introduction of disease vectors and exotics	(Depends on the situation)
Medium	Predator control	Medium
Low	Control of competitors	Low
Low	Control of prey species	Low

The above order of feasibility contains a high degree of subjectivity with obvious exceptions, but is based on the conclusion that single species manipulations, with the exception of exotics, whose introduction in some marine ecosystems has had dramatic (and usually adverse effects), are more uncertain in their effectiveness than manipulation of critical habitats, and of substrate species (e.g., coral reefs, mangroves and grassbeds). By the same token, it is considered that habitat protection or habitat improvement is more likely to be effective in marine ecosystem management than direct intervention in food webs by single-species control of the type mentioned lower in the above list. Critical habitat protection has the further advantage that it requires less detailed knowledge of species interactions, and at least preserves the options for future more sophisticated interventions in the food web that may appear feasible in the future.

22. REFERENCES

ACMRR/IABO Working Party on Ecological Indices of Stress to Fishery Resources, Indices for
1976 measuring responses of aquatic ecology to various human influences. A report of the ACMRR/IABO Working Party on ecological indices of stress to fishery resources. FAO Fish.Tech.Pap., (151):66 p.

Allen, K.R., Relation between production and biomass. J.Fish.Res.Board Can., 28(10):1573-81
1971

Allen, P.M. and J.M. McGlade, Dynamics of discovery and exploitation: the case of the Scotian
1986 Shelf groundfish fisheries. Can.J.Fish.Aquat.Sci., 43(6):1187-200

Alvarino, A., The relation between the distribution of zooplankton predators and anchovy larvae.
1980 Rep.CCOFI, (21):150-60

Alverson, D.L. and M.J. Carney, A graphic review of the growth and decay of population cohorts.
1975 J.Cons.CIEM, 36(2):133-43

Amaratunga, T., The role of cephalopods in the marine ecosystem. In Advances in assessment of
1983 world cephalopod resources, edited by J.F. Caddy. FAO Fish.Tech.Pap., (231):452 p.

Anderson, J.M., Ecology for environmental sciences: biosphere, ecosystems and man. New York,
1981   Halstead Press, Resource and environmental science series, 195 p.

Anderson, K.P. and E. Ursin, A multispecies extension to the Beverton and Holt theory of fishing,
1977   with accounts of phosphorus circulation and primary production. Medd.Danm.Fisk. Havunders., 7:319-435

Armstrong, M.J., The predator-prey relationships of Irish Sea poor-cod (Trisopterus minutus L.),
1982   pouting (Trisopterus luscus L.) and cod (Gadus morhua L.). J.Cons.CIEM, 40(2):135-52

Arnold, G.P., Squid: a review of their biology and fisheries. Lab.Leafl.Dir.Fish.Res.G.B.,
1979   (48):39 p.

Bacon, P.R., Overview of mollusc resources and exploitation in the Caribbean region. Paper
1984   presented at the 37th Meeting of the Gulf and Caribbean Fisheries Institute, and the Ninth Mexus-Gulf Cooperative Research Conference, Caneun, Mexico, November 1984 (mimeo)

Bagenal, T. (ed), Methods for assessment of fish production in fresh waters. IBP Handb., (3):
1978   365 p. 3rd ed.

Baisre, J.A., Los complejos ecológicos de pesca: definición e importancia en la administración de
1985   las pesquerías cubanas. FAO Fish.Rep./FAO Rapp.Pêches/FAO Inf.Pesca, (278) Suppl.:251-72

Bajkov, A.D., How to estimate the daily food consumption of fish under natural conditions. Trans.
1935   Am.Fish.Soc., 45:288-9

Bakun, A. and R.H. Parrish, Environmental inputs to fishery population models for eastern boundary
1980   current regions. IOC Workshop Rep., (28):67-104

Bakun, A., et al., Ocean sciences in relation to living resources. Can.J.Fish.Aquat.Sci., 39(7):
1980   1059-70

Baronov, F.I., On the question of the biological basis of fisheries. Izv.Nauchno.-Issled.Ikhtiol.
1918   Inst., 1:81-128 (in Russian)

Barrett, G.W., G.M. Van Dyne and E.P. Odum, Stress ecology. BioScience, 26(3):192-4
1976

Bas, G., R. Margalef and P. Rubies (eds), Simposio internacional sobre las áreas de afloramiento
1985   más importantes del oeste africano (cabo Blanco y Benguela). Barcelona, Instituto de Investigaciones Pesqueras, 2 vols:1114 p.

Beverton, R.J.H., Manual of methods for fish stock assessment. Part 2. Tables of yield functions
1966   for fishery assessment. Manuel sur les méthodes d'évaluation des stocks ichtyologiques. Partie 2. Tables des fonctions de rendement. Manual de métodos para la evaluación de los stocks de peces. Parte 2. Tablas de funciones de rendimiento. FAO Fish.Tech.Pap./FAO Doc.Tech.Pêches/FAO Doc.Téc.Pesca, (38):Rev.1:67 p.

Beverton, R.J.H. and S.J. Holt, On the dynamics of exploited fish populations. Fish.Invest.
1957   Minist.Agric.Fish.Food G.B.(2 Sea Fish.), 19:533 p.

_____, A review of the life spans and mortality rate of fish in nature, and their relation
1959   to growth and other physiological characteristics. In CIBA Foundation Colloquium on ageing, edited by G.E.W. Wolstenholme and M. O'Connor. London, Churchill, Vol.5:142-81

Beyer, J.E., Aquatic ecosystems: an operational research approach. Seattle, University of
1981   Washington Press, 315 p.

Beyer, J.E. and G.C. Laurence, 1981. Aspects of stochasticity in modelling growth and mortality
1981   of clupeoid fish larvae. Rapp.P.-V.Réun CIEM, 178:17-23

Blackburn, M., Micronekton of the eastern tropical Pacific Ocean: family composition, distri-
1968   bution, abundance, and relations to tuna. Fish.Bull.U.S.Fish.Wildl.Serv., 67(1):71-115

Blauenweiss, L. et al., Relationship between body size and some life history parameters.
1978   Oecologia, 37:257-72

Borgmann, U., Effect of somatic growth and reproduction on biomass transfer up pelagic food webs,
1983   as calculated from particle-size-conversion efficiency. Can.J.Fish.Aquat.Sci., 40(11):
        2010-8

Bradbury, R.H., R.E. Reichelt and D.G. Green,   Fractals in ecology: methods and interpretation.
1984   Mar.Ecol.(Prog.Ser.), 14(2-3):295-6

Breen, P.A. and K.H. Mann, Changing lobster abundance and the destruction of kelp beds by sea
1976   urchins. Mar.Biol., 34:137-42

Brett, J.R., Environmental factors and growth. In Fish physiology. Vol.8. Bioenergetics and
1979   growth, edited by W.S. Hoar, D.J. Randall and J.R. Brett. New York, Academic Press,
        pp. 599-675

Brett, J.R. and T.D.D. Groves, Physiological energetics. In Fish physiology. Vol.8. Bio-
1979   energetics and growth, edited by W.S. Hoar, D.J. Randall and J.R. Brett. New York,
        Academic Press, pp. 280-352

Butler, M.J.A and D. Jayasingh, The application of remote sensing technology to marine fisheries:
        an introductory manual. FAO Fish.Tech.Pap. (tentative title)(in preparation)

Butler, M.J.A., et al., Marine resource mapping: an introductory manual. FAO Fish.Tech.Pap.,
        (tentative title)(in preparation)

Caddy, J.F., Spatial model for an exploited shellfish population and its application to the
1975   Georges Bank scallop fishery. J.Fish.Res.Board Can., 32(8):1305-28

_____, Surplus production models. In Selected lectures from the CIDA/FAO/CECAF Seminar
1980   on fishery resource evaluation. Casablanca, Morocco, 6-24 March 1978. Rome, FAO,
        Canada Funds-in-Trust, FAO/TF/INT 180(c) (CAN) Suppl., pp. 29-55. Issued also in French

_____, Some considerations relevant to the definition of shared stocks and their allocation
1982   between adjacent economic zones. FAO Fish.Circ., (749):40 p. Issued also in Spanish

_____, The cephalopods: factors relevant to their population dynamics and to the assess-
1983   ment and management of stocks. FAO Fish.Tech.Pap., (231):419-52

_____, An alternative to equilibrium theory for management of fisheries. FAO Fish.Rep.,
1984   (289)Suppl. 2:173-214

_____, Species interactions and stock assessment - some ideas and approaches. In
1985   Simposio Internacional sobre las áreas de afloramiento más importantes del Oeste
        Africano (Cabo Blanco y Benguela), Barcelona, Instituto de Investigaciones Pesqueras,
        edited by C. Bas, R. Margalef and S.P. Rubies

_____, Modelling stock-recruitment processes in crustacea: some practical and theoretical
        perspectives. Can.J.Fish.Aquat.Sci., 44 (Suppl.)(in press)

Caddy, J.F. and G.P. Bazigos, Practical guidelines for statistical monitoring of fisheries in
1985   manpower-limited situations. FAO Fish.Tech.Pap., (257):86 p.

Caddy, J.F. and J. Csirke, Approximations to sustainable yield for exploited and unexploited
1983   stocks. Oceanogr.Trop., 18(1):3-15

Caddy, J.F. and S. Garcia, Production modelling without long data series. FAO Fish.Rep./FAO
1983   Rapp.Pêches/FAO Inf.Pesca, (268)Suppl.:309-13

_____, Fisheries thematic mapping : a prerequisite for intelligent management and de-
        velopment of fisheries. (in preparation)

Caddy, J.F. and J.A. Gulland, Historical patterns of fish stocks. Mar.Policy, 7(4):267-78
1983

Caddy, J.F. and T.D. Iles, Underwater observations on herring spawning grounds on Georges Bank.
1973   ICNAF Res.Bull., (10):131-9

Caddy, J.F. and G.D. Sharp, Some useful ecological concepts in fisheries assessment. Paper pre-
1983   sented at the ACMRR Working Party on the management of living resources in near-shore
        tropical waters. Rome, 28 February - 4 March, 1983, Working paper (8) (mimeo)

Carey, F.G. and R.J. Olson, Sonic tracking experiments with tunas. Collect.Vol.Sci.Pap.ICCAT/
1982   Rech.Doc.Sci.CICTA/Colecc.Doc.Sci.CICAA, 17:458-66

Carlenton, R.G., Man and the sea: the ecological challenge. Am.Zool., 25:451-68
1985

Chapman, A.R.O., Stability of sea urchin dominated barren grounds following destructive grazing of
1981   kelp in St Margarets Bay, eastern Canada. Mar.Biol., 61:307-11

Christy, F.T., Jr., Territorial user rights in marine fisheries: definitions and conditions. FAO
1982   Fish.Tech.Pap., (227):10 p. Issued also in French and Spanish

Christensen, B., Primary production of mangrove forests. In Proceedings of International
1978   Workshop for Mangrove and estuarine area development for the Indo-Pacific region.
        14-19 November 1977, Manila. Los Baños, Laguna, Philippine Council for Agriculture and
        Resources Research, pp. 131-6

Coelho, M.L. and A.A. Rosenberg, Causal analysis of some biological data for Illex illecebrosus
1984   from the Scotian Shelf. NAFO Sci.Counc.Stud., (7):61-6

Condrey, R.E., Density-dependent searching time: implications in surplus production models. Fish
1984   Bull.NOAA/NMFS, 82(3):537-42

Conover, R.J., Transformation of organic matter. In Marine ecology. Vol.4. Dynamics, edited by
1978   O. Kinne. Chichester, U.K., John Wiley and Sons, pp. 221-456

Conrad, M., Statistical and hierarchical aspects of biological organization. In Towards a
1972   theoretical biology, edited by C.H. Waddington. Edinburgh, Edinburgh University Press,
        Vol.4:189-221

_____, Biological adaptability: the statistical state model. Bioscience, 26(5):319-24
1976

Coulter, G.W., Population changes within a group of fish species in Lake Tanganyika following
1970   their exploitation. J.Fish Biol., 2:329-53

Cousins, S., Ecologists build pyramids again. New Sci., 106(1463):50-4
1985

Csirke, J., Recruitment in the Peruvian anchovy (Engraulis ringens) and its dependence on the
1980   adult population. Rapp.P.-V.Réun.CIEM, 177:307-13

Cushing, D.H., Climate and fisheries. London, Academic Press, 373 p.
1982

Cushing, D.H. and R.R. Dickson, The biological response in the sea to climatic changes. Adv.Mar.
1976   Biol., 14:1-122

Davies, R.G., Computer programming in quantitative biology. New York, Academic Press, 492 p.
1971

Day, D.S. and W.G. Pearcy, Species associations of benthic fishes on the continental shelf and
1968   slope of Oregon. J.Fish.Res.Board Can., 25(12):2665-75

De Sylva, D.P., Nektonic food webs in estuaries. In Estuarine research, edited by L.E. Cronin.
1975   New York, Academic press, pp. 420-47

Devries, T.J. and W.G. Pearcy, Fish debris in sediments of the upwelling zone off central Peru: a
1982   late Quaternary record. Deep-Sea Res.(A Oceanogr.Res.Pap.), 28(1):87-109

Diana, J.S., The feeding pattern and daily ration of a top carnivore, the northern pike (Esox
1979   lucius). Can.J.Zool., 57(11):2121-7

Dickie, L.M., Food chains and fish production. ICNAF Spec.Publ., (8):201-21
1972

Domain, F., Poissons démersaux du plateau continental Sénégambien: application de l'analyse aux
1972   composantes principales à une série de chalutages. Cah.ORSTOM (Sér.Océanogr.), 1(2):
        111-23

Dotson, R.C., Fat deposition and utilization in albacore. In The physiological ecology of
1978   tunas, edited by G.D. Sharp and A.E. Dizon. New York, Academic Press, pp. 343-55

Doubleday, W.G., Environmental fluctuations and fisheries management. Sel.Pap.ICNAF, (1)1:141-50
1976

Eggers, D.M., Factors in interpreting data obtained by diel sampling of fish stomachs. J.Fish.
1977   Res.Board Can., 34(2):290-4

Elliott, J.M., Some methods for the statistical analysis of samples of benthic invertebrates.
1971   Sci.Publ.Freshwat.Biol.Assoc.Ambleside, (25):144 p.

Elliott, J.M. and L. Persson, The estimation of daily rates of food consumption for fish. J.Anim.
1978   Ecol., 47:977-91

FAO, Yearbook of fishery statistics. Annuaire statistique des pêches. Anuario estadístico de
1984   pesca, 1983. Catches and landings. Captures et quantités débarquées. Capturas y
        desembarques. Yearb.Fish.Stat./Annu.Stat.Pêches/Anu.Estad.Pesca, (56):72

Fänge, R. and D. Grove, Digestion. In Fish physiology. Vol.8. Bioenergetics and growth, edited
1979   by W.S. Hoar, D.J. Randall and J.R. Brett. New York, Academic Press, pp. 161-260

Fasham, M.J.R., The statistical and mathematical analysis of plankton patchiness. Oceanogr.Mar.
1978   Biol., 16:43-79

_____, (ed.), Flows of energy and materials in marine ecosystems: theory and practice.
1984   New York, Plenum Press, NATO Conference series, IV, Marine sciences, 13:733 p.

Fisher, R.A., The design of experiments. Edinburgh, Oliver and Boyd, 7th ed.
1960

Fortunatova, K.R. and O.A. Popova, Feeding and food relationships of predatory fish in Volga
1973   delta. Moscow, Nauka Press

Fry, F.E.J., The effects of environmental factors on the physiology of fish. In Fish physiology,
1971   edited by W.S. Hoar and D.J. Randall. New York, Academic Press, Vol.6:1-98

Fuller, K.H., Shoreline as a controlling factor in commercial shrimp production. NASA Tech.Memo.,
1979   (72-73):227 p.

Gabriel, W.L. and A.V. Tyler, Preliminary analysis of Pacific coast demersal fish assemblages.
1980   Mar.Fish.Rev., 42(3-4):83-8

Gallucci, V.F., On the principles of thermodynamics in ecology. Annu.Rev.Ecol.Syst., 4:329-57
1973

Garcia, S., Distribution, migration and spawning of the main fish resources in the northern CECAF
1982   area. CECAF/ECAF Ser., (82/25):9 p.

_____, The stock-recruitment relationship in shrimps: reality or artefacts and misinter-
1983   pretations? Oceanogr.Trop., 18(1):25-48

Garcia, S. and L. Le Reste, Life cycles, dynamics, exploitation and management of coastal penaeid
1981   shrimp stocks. FAO Fish.Tech.Pap., (203):215 p. Issued also in French

Gilliland, R.L., Solar, volcanic and CO_2 forcing of recent climatic changes. Climat.Change,
1982   4:111-31

Goeden, G.B., Intensive fishing and a "keystone" predator species: ingredients for community
1982   instability. Biol.Conserv., 22:273-81

Golley, F.B., Energy flux in ecosystems. In Ecosystem structure and function, edited by
1972   J.A. Wiens. Proceedings of the Thirty-First Annual Biology Colloquium. Corvallis,
        Oregon State University Press, pp. 69-89

Grady, J.R., The distribution of sediment properties and shrimp catch on two shrimping grounds on
1971   the continental shelf of Mexico. Proc.Gulf Caribb.Fish.Inst., 23:39-48

Graham, M., (ed.), Sea fisheries: their investigation in the United Kingdom. London, Edward
1956   Arnold (Publishers), 487 p.

Grainger, R.J.R., Irish west coast herring fluctuations and their relation to oceanographic con-
1980   ditions. Rapp.P.-V.Réun.CIEM, 177:444 (Abstr.)

Greig-Smith, P., Quantitative plant ecology. London, Butterworths, 256 p. 2nd ed.
1964

Gulland, J.A., Food chain studies and some problems in world fisheries. In Marine food chains,
1970   edited by J.H. Steele. Edinburgh, Oliver and Boyd, pp. 296-315

_____, (comp.), The fish resources of the ocean. West Byfleet, Surrey, Fishing News
1971   (Books) Ltd., for FAO, 225 p. Rev.ed. of FAO Fish.Tech.Pap., (97):425 p.(1970)

Gulland, J.A. and L.K. Boerema, Scientific advice on catch levels. Fish.Bull.NOAA/NMFS, 71(2):
1973   32-5

Gulland, J.A. and S. Garcia, Observed patterns in multispecies fisheries. In Exploitation of
1984   marine communities edited by R.M. May. Report of the Dahlem Workshop on exploitation
        of marine communities. Berlin, 1-6 April 1984. Berlin, Springer-Verlag, Life sciences
        research report, 32:155-90

Gunderson, D.R., Using r-K selection theory to predict natural mortality. Can.J.Fish.Aquat.Sci.,
1980   37(12):2266-71

Hackney, P.A. and C.K. Minns, A computer model of biomass dynamics and food competition with
1974   implications for its use in fishery management. Trans.Am.Fish.Soc., 103(2):215-25

Hamblyn, E.L., The food and feeding habits of Nile perch Lates niloticus (Linne) (Pisces:
1966   Centroponidae). Rev.Zool.Bot.Afr., 74(1-2):28 p.

Hancock, D.A. and A.E. Urquhart, The determination of natural mortality and its causes in an
1965   exploited population of cockles (Cardium edule L.). Fish.Invest.Minist.Agric.Fish.Food
        G.B.(2 Sea Fish.), 24(2):40 p.

Hardy, A.C., On food and feeding habits of the herring. Fish.Invest.Minist.Agric.Fish.Food G.B.
1924   (2 Sea Fish.), 7(3)

Henderson, H.F., R.A. Ryder and A.W. Kudhongania, Assessing fishery potentials of lakes and
1973   reservoirs. J.Fish.Res.Board Can., 30(12) Part 2:2000-9

Hennemuth, R.C. and S.M. Autges, Effects of variability of recruitment on management service.
1982   ICES Pelagic Fish.Comm., CM 1982/H:22:4 p. (mimeo)

Hennemuth, R.C., J.E. Palmer and B.E. Brown, A statistical description of recruitment in eighteen
1980   selected fish stocks. J.Northwest Atl.Fish.Sci., 1:101-11

Heron, A.C., Population ecology of a colonizing species, the pelagic tunicate Thalia democratica.
1972   Part 1. Individual growth rate and generation time. Oecologia, 10(4):269-93

_____, Population ecology of a colonizing species, the pelagic tunicate Thalia democratica.
1973   Part 2. Population growth rate. Oecologia, 10(4):294-312

Hewitt, R.P., The value of pattern in the distribution of young fish. Rapp.P.-V.Réun.CIEM, 178:
1981   229-36

_____, Spatial pattern and survival of anchovy larvae: implications of adult reproductive
1982   strategy. Ph.D. Dissertation. University of California at San Diego, 187 p.

Hoar, W.S. and D.J. Randall (eds), Fish physiology. Vol.7. Locomotion. New York, Academic
1978   Press, 576 p.

Hoar, W.S., D.J. Randall and J.R. Brett (eds), Fish physiology. Vol.8. Bio-energetics and growth.
1979   New York, Academic Press, 786 p.

Hobson, E.S., Predatory behavior of some shore fishes in the Gulf of California. Res.Rep.U.S.Fish
1968   Wildl.Serv., (73):92 p.

_____, Diel feeding migrations in tropical reef fishes. Helgol.Wiss.Meeresunters., 24:
1973   361-70

Hobson, E.S., Feeding relationships of teleostean fishes on coral reefs in Kona, Hawaii. Fish.
1974   Bull.NOAA/NMFS, 72(4):915-1031

_____, Feeding patterns among tropical reef fishes. Am.Sci., 63:382-92
1975

Hobson, E.S. and J.R. Chess, Trophic interactions among fishes and zooplankters near shore at
1976   Santa Catalina Island, California. Fish.Bull.NOAA/NMFS, 74(3):567-98

_____, Trophic relationships among fishes and plankton in the lagoon at Enewetak Atoll,
1978   Marshall Islands. Fish.Bull.NOAA.NMFS, 76(1):133-53

Holden, M.J., Long-term changes in landing of fish from the North Sea. Rapp.P.-V.Réun.CIEM, 172:
1978   11-26

Holling, C.S., Resilience and stability of ecological systems. Annu.Rev.Ecol.Syst., 4:1-23
1973

Horne, E.P.W. and T. Platt, The dominant space and time scales of variablity in the physical and
1984   biological fields on continental shelves. Rapp.P.-V.Réun.CIEM, 138:8-19

Hughes, R.N. and M.L.H. Thomas, The classification and ordination of shallow-water benthic samples
1971   from Prince Edward Island. Can.J.Exp.Mar.Biol.Ecol., 7:1-39

Hunter, J.R. and R. Leong, The spawning energetics of female northern anchovy, Engraulis mordax.
1981   Fish.Bull.NOAA/NMFS, 79(2):215-30

Hurley, A.C., Feeding behaviour, food consumption, growth and respiration of the squid Loligo
1976   opalescens raised in the laboratory, Fish.Bull.NOAA/NMFS, 74(1):176-82

Huston, M., A general hypothesis of species diversity. Am.Nat., 13(1):81-101
1979

Hutchinson, B., The ecological theatre and the evolutionary play. New Haven, Yale University
1965   Press, 139 p.

Hutchinson, C.E., Concluding remarks. Cold Spring Harbour. Symp.Quant.Biol., 22:415-27
1957

Iles, T.D. and M. Sinclair, Atlantic herring: stock discreteness and abundance. Science, Wash.,
1982   215(4533):627-33

Isaacs, J.D., Unstructured marine food webs and "pollutant analogues". Fish.Bull.NOAA/NMFS,
1972   70(3):1053-9

_____, Potential trophic biomasses and trace substances concentrations in unstructured
1973   marine food webs. Mar.Biol., 22:97-104

_____, Reproductive products in marine food webs. Bull.South.Calif.Acad.Sci., (75):220-3
1976

Ivlev, V., Experimental ecology of the feeding of fishes. Transl. from Russian. New Haven,
1961   Connecticut, Yale University Press, 302 p.

Johannes, R.E., The metabolism of some coral reef communities a team study of nutrient and energy
1972   flux at Eniwetok. BioScience, 22(9):541-3

Johnson, L., The thermodynamic origin of ecosystems. Can.J.Fish.Aquat.Sci., 38(5):571-90
1981

Jones, R., Estimates of the food consumption of haddock (Melanogrammus aeglefinus) and cod (Gadus
1978   morhua). J.Cons.CIEM, 38:18-27

_____, The use of length composition data in fish stock assessment (with notes on VPA and
1981   cohort analysis). FAO Fish.Circ., (734):55 p. Issued also in French and Spanish

_____, Species interactions in the North Sea. Can.Spec.Publ.Fish.Aquat.Sci, (59):48-63
1982

Jones, R., Ecosystems, food chains and fish yields. ICLARM Conf.Proc., (9):195-239
1982a

_____, Assessing the effects of changes in exploitation pattern using length composition
1984   data (with notes on VPA and cohort analysis). FAO Fish.Tech.Pa., (256):118 p.

Jones, R. and E.W. Henderson, Further observations on energy flow through the marine food chain.
1980   ICES Biological Oceanography Cmtee. CM 1980/L:26 (mimeo)

Kapetsky, J.M., Some considerations for the management of coastal lagoon and estuarine fisheries.
1981   FAO Fish.Tech.Pap., (218):47 p. Issued also in French and Spanish

_____, Some potential environmental effects of coastal aquaculture with implications for
1982   site selection and aquaculture engineering. Manila, South China Sea Fisheries
        Development and Coordinating Programme, SCS/82/GEN/42:76-82

_____, Mangroves, fisheries and aquaculture. FAO Fish.Rep./FAO Rapp.Pêches/FAO Inf.
1985   Pesca, (338):Suppl.:17-36

Kawasaki, T., Fundamental relations among the selections of life history in the marine Teleost.
1980   Bull.Jap.Soc.Sci.Fish., 46(3):289-93

Kenchington, R.A. and B.E.T. Hudson (eds), Coral reef management handbook. Jakarta, Indonesia,
1984   Unesco Regional Office for Science and Technology for Southeast Asia, 281 p.

Kerr, S.R., Structural analysis of aquatic communities. In Proceedings of the First
1974   International Congress on ecology, The Hague. Wageningen, Netherlands Centre for
        Agricultural Publishing and Documentation

_____, Structure and transformation of fish production systems. J.Fish.Res.Board Can.,
1977   34(10):1989-93

_____, Niche theory in fisheries ecology. Trans.Am.Fish.Soc., 109:254-60
1980

_____, Estimating the energy budgets of actively predatory fish. Can.J.Fish.Aquat.Sci.,
1982   39(3):371-9

Kerr, S.R. and R.A. Ryder, Niche theory and percid community structure. J.Fish.Res.Board Can.,
1977   34(10):1952-58

Kinne, O., (ed.), Marine ecology. A comprehensive, integrated treatise on life in oceans and
1970-84 coastal waters. Chichester, U.K., John Wiley and Sons, Vol.1, Pts 1-3; Vol.2. Pts
        1-2; Vol.4; Vol.5, Pts 1-4

Knight, W. and A.V. Tyler, A method for compression of species association data by using habitat
1973   preferences, including an analysis of fish assemblages on the southwest Scotian Shelf.
        Tech.Rep.Fish.Res.Board.Can., (402):15 p.

Kondo, K., The recovery of the Japanese sardine - the biological basis of stock-size fluctuations.
1980   Rapp.P.-V.Réun.CIEM, 177:332-54

Laevastu, T. and F. Favorite, Holistic simulation models of shelf-seas ecosystems. In Analysis
1980   of marine ecosystems, edited by A.R. Longhurst. London, Academic Press, pp. 701-27

Laevastu, T. and H.A. Larkins, Marine fisheries ecosystem: its quantitative evaluation and
1981   management. Farnham, Surrey, Fishing News Books Ltd., 162 p.

Lamson, C. and A.J. Hanson (eds), Atlantic fisheries and coastal communities: fisheries decision-
1984   making case studies. Halifax, Nova Scotia, Dalhousie University, Institute for
        Resource and Environmental Studies, 262 p.

Lane, E.D., Quantitative aspects of the life history of the diamond turbot, Hypsopsetta gluttulata
1975   (Girard), in Anaheim Bay. Fish.Bull.Calif.Dep.Fish Game, (165):153-73

Lane, E.D., M.C.S. Kingsley and D.E. Thornton, Daily feeding and food conversion efficiency of the
1979   diamond turbot: an analysis based on field data. Trans.Am.Fish.Soc., 108:530-5

Larkin, P.A., Interspecific competition and exploitation. J.Fish.Res.Board Can., 20(3):647-78
1963

Larkin, P.A. and S.W. Gazey, Applications of ecological simulation models to management of
1982   tropical multispecies fisheries. ICLARM Conf.Proc., (9):123-40

Lasker, R., Feeding, growth, respiration and carbon utilization of a euphausid crustacean.
1966   J.Fish.Res.Board Can., 23(9):1291-317

_____, Ocean variability and its biological effects-regional review northeast Pacific.
1978   Rapp.P.-V.Réun.CIEM, 173:168-81

Lessios, H.A., P.W. Glynn and J.D. Cubit, Spread of Diadema mass mortality through the Caribbean.
1984   Science,Wash., 226:335-7

Li, H.W. and P.B. Moyle, Ecological analysis of species introductions into aquatic systems.
1981   Trans.Am.Fish.Soc., 110:772-82

Lie, U. and J.C. Kelley, Benthic infauna communities off the coast of Washington and in Puget
1970   Sound: identification and distribution of the communities. J.Fish.Res.Board Can., 27(1):621-51

Lindemann, R.L., The trophic-dynamic aspect of ecology. Ecology, 23:399-418
1942

Livingston, R.J., Trophic organization of fishes in a coastal seagrass system. Mar.Ecol.(Prog.
1982   Ser.), 7:1-12

Lleonart, J., J. Salet and E. MacPherson, Effect of the cannibalism in hake population. ICES
1983   Demersal Fish Committee CM 1983/G:54:22 p. (mimeo)

_____, Efecto del canibalismo en la población de Merluccius capensis en la División 1.5.
1983a   Collect.Sci.Pap.ICSEAF/Recl.Doc.Sci.CIPASE/Colecc.Doc.Cient.CIPASO, 10(1):111-28

Longhurst, A.R., Analysis of marine ecosystems. New York, Academic Press, 741 p.
1981

Louins, A.B. and L.H. Louins, Energy: what is the problem? Ecologist, 11(6):302-13
1981

MacArthur, R., Fluctuations of animal populations and a measure of community stability. Ecology,
1955   36(3):533-6

MacArthur, R.H. and E.O. Wilson, The theory of island biogeography. Princeton, N.J., Princeton
1967   University Press, 25 p.

MacDonald, J.S., et al., Fishes, fish assemblages and their seasonal movements in the lower Bay of
1984   Fundy and Passamaquoddy Bay, Canada. Fish.Bull.NOAA/NMFS, 82(1):121-39

MacDonald, N., Trees and networks in biological models. New York, John Wiley, 216 p.
1983

Mackinnon, J.C., Analysis of energy flow and production in an unexploited marine flatfish popula-
1973   tion. J.Fish.Res.Board Can., 30(11):717-28

_____, Metabolism and its relationship with growth rate of American plaice, Hippoglossoides
1973a   platessoides Fabr. J.Exp.Mar.Biol.Ecol., 11:297-310

MacPherson, E., Resource partitioning in a Mediterranean demersal fish community. Mar.Ecol.(Prog.
1981   Ser.), 4(2):183-94

Magnuson, J.J., Digestion and food consumption by skipjack tuna (Katsuwonis pelamis). Trans.Am.
1969   Fish.Soc., 98:379-92

Magnuson, J.J. and J.G. Heitz, Gill raker apparatus and food selectivity among mackerels, tunas
1971   and dolphins. Fish.Bull.NOAA/NMFS, 69(2):361-70

Mahon, R., H. Oxenford and W. Hunte (eds), Development strategies for flying fish fisheries of the
1986   eastern Caribbean. Proceedings of an IDRC-sponsored Workshop at the University of the
        West Indies, Cave Hill, Barbados, 22-23 October 1985. Manuscr.Rep.Int.Dev.Res.Cent.,
        Ottawa, (IDRC-MR128e):147 p.

Mahon, R., et al., Spatial and temporal patterns of groundfish distribution on the Scotian Shelf
1984 and in the Bay of Fundy, 1970-1981. Can.Tech.Rep.Fish.Aquat.Sci., (1300):164 p.

Mandelbrot, B.B., The fractal geometry of nature. New York, W.H. Freeman, 460 p.
1982

Mann, K.H., Ecological energetics of the seaweed zone in a marine bay on the Atlantic coast of
1972 Canada. 2. Productivity of the seaweeds. Mar.Biol., 14:199-209

_____, Destruction of kelp-beds by sea urchins: a cyclic phenomenon or irreversible
1977 degradation? Helgol.Wiss.Meeresunters., 30:455-67

_____, Estimating the food consumption of fish in nature. In Ecology of freshwater fish
1978 production, edited by S.D. Gerking. Edinburgh, Oxford Blackwell Scientific
Publications, pp. 250-73

_____, Ecology of coastal waters: a systems approach. Berkeley, California, University
1982 of California Press, 322 p.

Margalef, R., Perspectives in ecological theory. Chicago, University of Chicago Press, 111 p.
1968

_____, General concepts of population dynamics and food links. In Marine ecology. Vol.4.
1978 Dynamics, edited by O. Kinne. Chichester, U.K., Wiley-Interscience, pp. 617-704

Mark, D.M., Fractal dimension of a coral reef at ecological scales: a discussion. Mar.Ecol.
1984 (Prog.Ser.), 14:293-4

Marten, G.C. and J.J. Polovina, A comparative study of fish yields from various tropical
1982 ecosystems. ICLARM Conf.Proc., 9:255-85

Martosubroto, P. and N. Naamin, Relationship between tidal forests (mangroves) and commercial
1977 shrimp production in Indonesia. Mar.Res.Indones., 18:81-6

May, R.M., Stability and complexity in model ecosystems. Princeton, N.J., Princeton University
1974 Press. Princeton N.J.

_____, (ed.) Exploitation of marine communities. Report of the Dahlem Workshop on
1984 exploitation of marine communities. Berlin 1984, April 1-6. Berlin, Springer-Verlag,
Dahlem Workshop reports, Life sciences research report, 32:366 p.

May, R.M., et al., Management of multispecies fisheries. Science, Wash., 205(4403):267-77
1979

McNeill, S. and J.H. Lawton, Annual production and respiration in animal populations. Nature,
1970 Lond., 225(5231):472-4

Mearns, A.J., et al., Trophic structure and the caesium-potassium ratio in pelagic ecosystems.
1981 Rep.CCOFI, (22):99-110

Miller, R.B., A succession in sea urchin and seaweed abundance in Nova Scotia, Canada. Mar.Biol.,
1985 84:275-86

_____, Sea urchin pathogen: a possible tool for biological control. Mar.Ecol.(Prog.Ser.),
1985a 21:169-74

Miller, R.J., K.H. Mann and D.J. Scarratt, Production potential of a seaweed lobster community in
1971 eastern Canada. J.Fish.Res.Board Can., 28(11):1733-8

Mills, E.L., K. Pittman and F.C. Tan, Food web structure on the Scotian Shelf, eastern Canada: a
1982 study using 13C as a food-chain tracer. Paper presented at Symposium on Biological
productivity of continental shelves in the temperate zone of the north atlantic. Kiel,
F.R.G., 2-5 March 1982, Doc. No. 29 (mimeo)

Morse, D.R., et al., Fractal dimension of vegetation and the distribution of arthropod body
1985 lengths. Nature,Lond., 314:731-3

Munro, J.L., The food of a community of East African fish. J.Zool.Lond., 151:389-415
1967

Munro, J.L., Actual and potential fish production from the coralline shelves of the Caribbean
1977 Sea. FAO Fish.Rep., (200):301-21

_____, Stock assessment models: applicability and utility in tropical small-scale
1979 fisheries. In Stock assessments for tropical small-scale fisheries. Proceedings of the International Workshop, University of Rhode Island, 19-21 September 1979, edited by S.B. Saila and P.M. Roedel. Kingston, R.I., University of Rhode Island Press, International Center for marine Resource Development, pp. 35-47

_____, Estimation of biological and fishery parameters in coral reef fisheries. ICLARM
1982 Conf.Proc., (9):71-82

_____, (ed.), Caribbean coral reef fishery resources. ICLARM Stud.Rev., (7):276 p.
1983

_____, Yields from coral reef fisheries. ICLARM Fishbyte, 2(3):13-5
1984

Murdoch, W.W., Switching in general predators: experiments on predator specificity and stability
1969 of prey populations. Ecol.Monogr., 39(4):335-54

Newell, R.C., The biological role of detritus in the marine environment. In Flows of energy and
1984 materials in marine ecosystems: theory and practice edited by M.J.R. Fasham. New York, Plenum Press, NATO Conference Series, IV, Marine sciences, Vol. 13, pp. 317-43

Nicolis, G. and I. Prigogine, Self-organization in non-equilibrium: to dissipative structures to
1977 order through fluctuations. New York, Wiley-Interscience, 491 p.

Nixon, S.W., Nutrient dynamics, primary production and fisheries yields of lagoons. Oceanol.Acta,
1982 5(4)Suppl. pp.357-71

Norman, J.R., A history of fishes. London, Ernest Benn Ltd., 398 p. 2nd ed. Edited by
1963 P.H. Greenwood

Odum, E.P., Energy flow in ecosystems: a historical review. Am.Zool., 8:217-24
1968

_____, The strategy of ecosystems development. Science,Wash., 164:262-70
1969

Odum, H.T., Biological circuits and the marine systems of Texas. In Pollution and marine ecology,
1967 edited by T.A. Olson and F.J. Burgess. New York, Wiley Interscience, pp. 99-157

_____, An energy circuit language for ecological and social systems: its physical basis.
1972 In System analysis and simulations in ecology, edited by B.C. Patten. New York, Academic Press, pp. 140-211

Olson, R.J., Feeding and energetics studies of yellowfin tuna; food for ecological thought.
1982 Collect.Vol.Sci.Pap.ICCAT/Rech.Doc.Sci.CICTA/Colecc.Doc.Cient.CICAA, 17:444-57

Overholtz, W.J. and A.V. Tyler, Long-term responses of the demersal fish assemblages of Georges
1985 Bank. Fish.Bull.NOAA/NMFS, 83(4):507-20

Owen, R.W., Patterning of flow and organisms in the larval anchovy environment. IOC Workshop
1980 Rep., (28):1167-200

Paine, R.T., The Pisaster-Tegula interaction: prey patches, predator food preference and inter-
1969 tidal community structure. Ecology, 50:930-8

_____, A note on trophic complexity and community stability. Am.Nat., 103:91-3
1969a

Paloheimo, J.E., Studies on estimation of mortalities. 1. Comparison of a method described by
1961 Beverton and Holt and a new linear formula. J.Fish.Res.Board Can., 18(5):645-62

Paloheimo, J.E. and L.M. Dickie, Production and food supply. In Marine food chains, edited by
1970 J.H. Steele. Edinburgh, Oliver and Boyd, pp. 499-527

Paloheimo, J.E. and H.A. Regier, Ecological approaches to stressed multispecies fisheries re-
1982 sources. Can.Spec.Publ.Fish.Aquat.Sci., (59):127-32

Parrish, R.H. and A.D. MacCall, Climatic variation and exploitation in the Pacific mackerel
1978   fishery. Fish Bull.Cal.Dep.Fish Game, (167):110 p.

Parrish, R.H., C.S. Nelson and A. Bakun, Transport mechanisms and reproductive success of fishes
1981   in the California Current. Biol.Oceanogr., 1(2):175-203

Parrish, J.D. and R.J. Zimmerman, Utilization by fishes of space and food resources on an offshore
1977   Puerto Rican coral reef and its surrounding. In Proceedings of the third International
        Coral Reef Symposium. Miami, Florida, University of Miami, Rosentiel School of Marine
        and Atmospheric Science, Vol.1:297-303

Passet, R., L'économique et le vivant. Paris, Payot, Petite Bibliothèque, 286 p.
1979

Pauly, D., Theory and management of tropical multispecies stocks: a review, with emphasis on the
1979   Southeast Asian demersal fisheries. ICLARM Stud.Rev., (1):35 p.

_____, Gill size and temperature as governing factors in fish growth: a generalization of
1979a   the von Bertalanffy growth formula. Ber.Inst.Meereskd.Christian-Albrechts Univ.Kiel,
        (63)

_____, A selection of simple methods for the assessment of tropical fish stocks. FAO Fish.
1980   Circ., (729):54 p. Issued also in French.

_____, On the interrelationships between natural mortality, growth parameters and mean en
1980a   vironmental temperature in 175 fish stocks. J.Cons.CIEM, 39(3):195-212

_____, Fish stock assessment in coral reefs - notes on the state of the art. ICLARM Newsl.,
1981   4(3):19

_____, Some simple methods for the assessment of tropical fish stocks. FAO Fish.Tech.Pap.,
1983   (234):52 p. Issued also in French and Spanish. Revision of FAO Fish.Circ.,
        (729):54 p.

_____, Fish population dynamics in tropical waters: a manual with application programs for
1984   programmable calculators. ICLARM Stud.Rev., (8):325 p.

Pauly, D. and A.N. Mines, (eds), Small-scale fisheries of San Miguel Bay, Philippines: biology
1982   and stock assessment. ICLARM Tech.Rep., (7):124 p.

Pauly, D. and G.I. Murphy (eds), Theory and management of tropical fisheries. Proceedings of the
1982   ICLARM/CSIRO Workshop on the theory and management of tropical multispecies stocks.
        12-21 January 1981, Cronulla, Australia. ICLARM Conf.Proc., (9):360 p.

Peterson, I. and J.S. Wroblewski, Mortality rate of fishes in the pelagic ecosystem. Can.J.Fish.
1984   Aquat.Sci., 41(7):1117-20

Pianka, E.R., On r-selection and K-selection. Am.Nat., 104:592-7
1970

_____, r and K - selection or b and d selection? Am.Nat., 106:581-8
1972

Pielou, E.C., An introduction to mathematical ecology. New York, Wiley Interscience, 286 p.
1969

Pimm, S.L., Properties of food webs. Ecology, 61(2):219-25
1980

Pinkas, L., M.S. Oliphant and I.L.K. Iverson, Food habits of albacore, bluefin tuna and bonito in
1971   California waters. Fish Bull.Calif.Dep.Fish Game, (152):1-105

Pitt, T.K., Estimates of annual mortality coefficients of American plaice. ICNAF Res.Doc.,
1972   (72/15) Serial No. (2699):12 p.

_____, American plaice from ICNAF subarea 2 and divisions 3K, a stock assessment update.
1978   CAFSAC Res.Doc./CSPA Doc.Rech., Dartmouth, N.S., (78/11):6 p. (mimeo)

Platt, T. and K. Denman, A general equation for the mesoscale distribution of phytoplankton in the
1975 sea. Mem.Soc.R.Sci.Liege(60 Ser.), 7:31-42

_____, Organization in the pelagic ecosystem. Helgol.Wiss.Meersunters., 30:575-81
1977

_____, The structure of pelagic marine ecosystems. Rapp.P.-V.Réun.CIEM, 173:60-5
1978

Platt, T., K.H. Mann and R.E. Ulanowicz (eds), Mathematical models in biological oceanography.
1981 Monogr.Oceanogr.Methodol., (7):156 p.

Pollard, D.A., A review of ecological studies on seagrass - fish communities, with particular
1984 reference to recent studies in Australia. Aquat.Bot., 18:3-42

Polovina, J.J. and M.D. Ow, An approach to estimating an ecosystem box model. Fish.Bull.NOAA/NMFS,
1985 83(3):457-60

Pomeroy, L.R., The oceans' food web, a changing paradigm. Bioscience, 24(9):499-504
1974

Pope, J.G., A modified cohort analysis in which constant natural mortality is replaced by
1979 estimates of predation level. ICES CM.1981/G:14 (mimeo)

_____, Phalanx analysis: an extension of Jones; length cohort analysis to multispecies
1980 cohort analysis. ICES Demersal Fish Committee. CM.1980/G:19:5 p. (mimeo)

_____, Assessment of multispecies resources. In Selected lectures from the CIDA/FAO/CECAF
1980a Seminar on fishery resource evaluation. Casablanca, Morocco, 6-24 March 1978. Rome,
FAO, Canada Funds-in-trust, FAO/TF/INT 180(c)(CAN)Suppl. Issued also in French

Pope, J.G. and L. Woolner, A simple investigation into the effects of predation on recruitment to
1981 some North Sea fish stocks. ICES CM.1981/G:14:4 p. (mimeo)

Popova, O.A., The role of predacious fish in ecosystems. In Ecology of freshwater fish produc-
1978 tion, edited by S.D. Gerking. Oxford, Blackwell Scientific Publications, pp. 215-49

Powles, H. and C.A. Barans, Groundfish monitoring in sponge-coral areas off the Southeastern
1980 United States. Mar.Fish.Rev., 42(5):21-35

Prigogine, I., Time structure and fluctuations. Science, Wash., 201(4358):777-85
1978

Pringle, J.D., G.J. Sharp and J.F. Caddy, Interactions in kelp bed ecosystems in the northwest
1982 Atlantic: review of a Workshop. Can.Spec.Publ.Fish.Aquat.Sci., (59):108-15

Ran, G.H., et al., Animal $^{13}C/^{12}C$ correlates with trophic level in pelagic food webs. Ecology,
1983 64(5):1314-8

Randall, J.E., Food habits of reef fishes of the West Indies. Stud.Trop.Oceanogr., 5:665-847
1967

Rapport, D.J., H.A. Regier and T.C. Hutchinson, Ecosystem behaviour under stress. Am.Nat., 125
1985 (5):617-38

Rasmusson, E.M., El Niño: the ocean/atmosphere connection. Oceanus, 27(2):5-12
1984

Regier, H.A., Sequence of exploitation of stocks in multispecies fisheries in the Laurentian Great
1973 Lakes. J.Fish.Res.Board Can., 30(12)Pt.2:1992-9

Regier, H.A. and H.F. Henderson, Towards a broad ecological model of fish communities and
1973 fisheries. Trans.Am.Fish.Soc., 102(1):56-72

Regier, H.A. and K.H. Loftus, The ecological role of fisheries exploitation in salmonid commu-
1972 nities. J.Fish.Res.Board Can., 29(7):959-68

Reintjes, J.W. and J.E. King., Food of yellowfin tuna in the Central Pacific. Fish.Bull U.S. Fish
1953 Wildl.Serv., 54:90-110

Rheinheimer, G., Aquatic microbiology. Chichester, U.K., John Wiley and Sons, 235 p. 2nd ed.
1980   Transl. from the German

Ricker, W.E., Handbook of computations for biological statistics of fish populations. Bull.Fish.
1958   Res.Board Can., (119):300 p.

———, Linear regression in fishery research. J.Fish.Res.Board Can., 30(3):409-34
1973

———, Computations and interpretation of biological statistics of fish populations. Bull.
1975   Fish.Res.Board Can., (191):382 p.

———, Growth rates and models. In Fish physiology. Vol.8. Bioenergetics and growth,
1979   edited by H.S. Hoar, D.J. Randall and J.R. Brett. New York, Academic Press,
       pp. 677-743

Rigler, F.H., The relation between fishery management and limnology. Trans.Am.Fish.Soc., 111(2):
1982   121-32

Rikhter, V.A. and V.N. Efanov, On one of the approaches to estimates of natural mortality of fish
1976   populations. ICNAF Res.Doc., 76(VI) 8:12 p.

Robson, D.S., Maximum likelihood estimation of a sequence of annual survival rates from a capture/
1963   recapture series. Spec.Publ.ICNAF, (4):330-5

Roger, C. and R. Grandperrin, Pelagic food webs in the tropical Pacific. Limnol.Oceanogr., 21:
1976   731-5

Rollet, B., La ecología de los manglares con referencia especial a la base biológica para la
1984   ordenación sostenida, forestal y pesca. Paper presented at the FAO Seminario regional
       sobre la ordenación integrada de los manglares. Cuba-Venezuela-Colombia, 5-23 November
       1984. 33 p. (mimeo)

Romesburg, H.C., Cluster analysis for researchers. Belmont, California, Lifetime Learning
1984   Publications

Rowe, G.T., Benthic biomass and surface productivity. In Fertility of the sea, edited by
1971   J.D. Costlow. New York, Gordon and Breach, vol.2:441-54

Ryther, J.H., Photosynthesis and fish production in the sea. Science,Wash., 166:72-6
1969

Saenger, P., E.J. Hegerl and J.D.S. Davie (eds), Global status of mangrove ecosystems. IUCN
1983   Comm.Ecol.Pap., (3):88 p.

Saila, S.B. and T.A. Gaucher, Estimation of the sampling distribution and numerical abundance of
1966   some mollusks in a Rhode Island Salt pond. Proc.Natl.Shellfish.Assoc., 56:73-80

Saila, S.B. and J.D. Parrish, Exploitation effects upon interspecific relationships in marine
1972   ecosystems. Fish.Bull.NOAA/NMFS, 70(2):383-93

Sainsbury, K.J., The ecological basis of tropical fisheries management. ICLARM Conf.Proc.,
1982   9:167-94

Sanders, M.J., Estimation of mortality coefficients for exploited fish populations in which the
1977   sexes exhibit divergent growth rates. J.Cons.CIEM., 37(2):192-3

Santander, H., Patrones de distribución y fluctuaciones de desoves de anchoveta y sardina. In
1981   Volumen extraordinario: Investigación cooperativa de la anchoveta y su ecosistema -
       ICANE-entre Perú y Canadá. Callao, Instituto del Mar del Perú

Saunders, P.T., An introduction to catastrophe theory. Cambridge, U.K., Cambridge University
1980   Press, 144 p.

Schaefer, M.B., G.C. Broadhead and C.J. Orange, Synopsis on the biology of yellowfin tuna, Thunnus
1963   (Neothunnus) albacares (Bonnaterre) 1788 (Pacific Ocean). FAO Fish.Rep., (6)vol.2:
       538-61

Schlesinger, D.A. and H.A. Regier, Climatic and morphoedaphic indices of fish yields from natural
1982   lakes. Trans.Am.Fish.Soc., 111:141-50

Seaburg, K.G. and J.B. Moyle, Feeding habits, digestion rates, and growth of some mennesota
1964   warmwater fishes. Trans.Am.Fish.Soc., 93(3):269-85

Selye, H., Stress without distress. Philadelphia, J.B. Lippincott Co., 171 p.
1974

Sharp, G.D., Behavioral and physiological properties of tuna and their effects on vulnerability to
1978   fishing gear. In The physiological ecology of tunas, edited by G.D. Sharp and
        A.E. Dizon. New York, Academic Press, pp. 397-450

_____, (Rapporteur), Report of the Workshop on the effects of environmental variation on
1980   survival of larval pelagic fishes. IOC Workshop Rep., (28):1-47

_____, Colonization: modes of opportunism in the ocean. IOC Workshop Rep., (28):125-48
1980a

_____, (ed.), Workshop on the effects of environmental variation on the arrival of larval
1980b   pelagic fishes. IOC Workshop Rep., (28):323 p.

_____, Tuna fisheries, elusive stock boundaries and the illusory stock concepts. Collect.
1983   Vol.Sci.Pap.ICCAT/Rech.Doc.Sci.CICTA/Colecc.Doc.Cient.CICAA, 18(3):812-29

_____, Ecological efficiency and activity metabolism. In Flows of energy and materials in
1983a   marine ecosystems: theory and practice, edited by M. Fasham. New York, Plenum Press,
        NATO Conference series, IV, Marine sciences, vol. 13

Sharp, G.D. and J. Csirke (eds), Proceedings of the Expert Consultation to examine changes in
1983   abundance and species of neritic fish resources. San José, Costa Rica, 18-29 April
        1983. Actas de la Consulta de Expertos para examinar los cambios en la abundancia y
        composición por especies de recursos de peces neríticos. San José, Costa Rica,
        18-29 abril 1983. A preparatory meeting for the FAO World Conference on fisheries
        management and development. Una reunión preparatoria para la Conferencia Mundial de la
        FAO sobre ordenación y desarrollo pesqueros. FAO Fish.Rep./FAO Inf.Pesca,
        (291)Vol.2:553 p. FAO Fish.Rep./FAO Inf.Pesca, (291):Vol.3:557-1224

Sharp, G.D., J. Csirke and S. Garcia, Modelling fisheries: what was the question? FAO Fish.Rep./
1983   FAO Inf.Pesca, (291)Vol.3:1177-224

Sharp, G.D. and R.C. Dotson, Energy for migration in albacore, Thunnus alalunga. Fish.Bull.NOAA/
1977   NMFS, 75(2):447-50

Sharp, G.D. and R.C. Francis, An energetics model of the exploited yellowfin tuna, Thunnus
1976   albacares, population in the eastern Pacific Ocean. Fish.Bull.NOAA/NMFS, 74(1):36-50

Sheldon, R.W. and T.R. Parsons, A continuous size spectrum for particulate matter in the sea.
1967   J.Fish.Res.Board Can., 24(5):909-15

Sheldon, R.W., A. Prakash and W.H. Sutcliffe, The size distribution of particles in the ocean.
1972   Limnol.Oceanogr., 17(3):327-40

Sheridan, P.F., J.A. Browder and J.E. Powers, Ecological interactions between penaeid shrimp and
1984   bottomfish assemblages. In Penaeid shrimps - their biology and management, edited by
        J.A. Gulland and B.J. Rothschild. Farnham, Surrey, Fishing News Books, Ltd.,
        pp. 235-53

Shirakihara, K. and S. Tanaka, A predator-prey model for two-species populations with nonlinear
1981   interactions and implications for fisheries management. Bull.Jap.Soc.Sci.Fish., 47(4):
        487-93

Silliman, R.P., Studies on the Pacific pilchard or sardine (Sardinops caerulea). 5. A method of
1943   computing mortalities and replacements. Spec.Sci.Rep.U.S.Fish Wildl.Serv., (24):10 p.

Simpson, J.H. et al., The Islay front: physical structure and phytoplankton distribution,
1979   estuarine and coastal. Mar.Sci., 9:713-26

Sinclair, M., Marine populations: an essay on population regulation and speciation in the oceans.
        Seattle University of Washington/US Sea Grant Publication (in press)

Sissenwine, M.P., Why do fish populations vary? In Exploitation of marine communities, edited by
1984   R.M. May. Berlin, Springer Verlag, Dahlem Workshop reports, Life sciences research reports, 32:59-94

Skud, B.E., Dominance in fishes: the relation between environment and abundance. Science, Wash.,
1982   216:144-9

Slobodkin, L.B., Growth and regulation of animal populations. New York, Holt, Rinehart and
1961   Winston, 184 p.

Smith, C.L., Analysis of a coral reef fish community: size and relative abundance. Hydro-lab J.,
1975   3:31-8

Smith, C.L. and J.C. Tyler, Population ecology of a Bahamian superbenthic shorefish assemblage.
1973   Am.Mus.Novit., (2528):38 p.

_____, Succession and stability in fish communities of dome-shaped patch reefs in the West
1975   Indies. Am.Mus.Novit., (2572):18 p.

Soutar, A. and J.D. Isaacs, Abundance of pelagic fish during the 19th and 20th centuries as
1974   recorded in anaerobic sediment of the Californias. Fish.Bull.NOAA/NMFS, 72(2):275-94

Southward, A.J., E.I. Butter and L. Pennycuik, Recent cyclic changes in climate and in abundance
1975   of marine life. Nature, Lond., 253:714-7

Steel, R.G. and J.H. Torrier, Principles and procedures of statistics: a biometrical approach.
1980   New York, McGraw-Hill Book Co. Ltd., 481 p. 2nd ed.

Steinbeck, J., The log from the "Sea of Cortez". London, William Heinemann Ltd. Republished in
1958   1974 by Pan Books Ltd., London, 320 p.

Stillwell, C.E. and N.E. Kohler, Food, feeding habits, and estimates of daily ration of the
1982   shortfin mako (Isurus oxyrinchus) in the northwest Atlantic. Can.J.Fish.Aquat.Sci., 39(3):407-14

Swartzman, G., Evaluation of ecological simulation models. In Contemporary quantitative ecology
1979   and related econometrics, edited by G.P. Patil and M. Rosenzweig. Fairland, Maryland, International Cooperative Publishing House, pp. 295-318

Swenson, W.A. and L.L. Smith Jr., Gastric digestion, food consumption, feeding periodicity, and
1973   food conversion efficiency in walleye (Stizostedion vitraeum vitreum). J.Fish.Res. Board Can., 30(9):1327-36

Taylor, C.C., Nature of variability in trawl catches. Fish.Bull.U.S.Fish Wildl.Serv., 54:145-66
1953

_____, Growth equations with metabolic parameters. J.Cons.CIEM, 27:270-86
1962

Thayer, C.W., S.M. Adams and M.W. La Croix, The importance of eelgrass beds in Puget Sound. Mar.
1975   Fish.Rev., 39:18-22

Theilacker, G. and K.T. Dorsey, Larval fish diversity, a summary of laboratory and field research.
1980   IOC Workshop Rep., (28):105-42

Thorhaug, A., Large-scale seagrass restoration in a damaged estuary. Mar.Pollut.Bull., 16(2):
1985   55-62

Thorpe, J.E., Daily ration of adult perch, Perca fluviatilis L., during summer in Loch Leven,
1977   Scotland. J.Fish Biol., 11:55-68

Thorson, G., Bottom communities (sublittoral or shallow shelf). Mem.Geol.Soc.Am., (67):461-534
1957

Tremblay, C. and F. Axelsen, Données sur la pêche, la biologie et l'abondance du flétan du
1980   Groënland (Reinhardtius hippoglossoides) dans le Golfe du Saint-Laurent. CAFSAC Res.Doc./CSCPA Doc.Rech., Dartmouth, N.S., (80/34):26 p.

Troadec, J.-P., W.G. Clark and J.A. Gulland, A review of some pelagic fish stocks in other areas.
1980   Rapp.P.-V. Réun.CIEM, 177:252-77

Tyler, A.V., Periodic and resident components in communities of Atlantic fishes. J.Fish.Res.
1971   Board Can., 28(7):935-46

Tyler, A.V., W.L. Gabriel and W.J. Overholtz, Adaptive management based on structure of fish
1982   assemblages of northern continental shelves. Can.Spec.Publ.Fish.Aquat.Sci., (59):
       149-56

Ulanowicz, R.E., An hypothesis on the development of natural communities. J.Theor.Biol.,
1980   85:223-45

Ursin, E., Stability and variability in the marine ecosystem. Dana, 2:51-67
1982

Vandermeer, J., Elementary mathematical ecology. New York, John Wiley, 294 p.
1981

Vinogradov, M.E., Feeding of the deep-sea zooplankton. Rapp.P.-V.Réun.Cons.CIEM, 153:114-20
1962

Vlymen, W.J., Energy expenditure of swimming copepods. Limnol.Oceanogr., 15(3):348-56
1970

_____, Swimming energetics of the larval anchovy, Engraulis mordax. Fish.Bull.NOAA/NMFS,
1974   72(4):885-99

_____, A mathematical model of the relationship between larval anchovy (E. mordax) growth,
1977   prey microdistribution and larval behaviour. Environ.Biol.Fishes, 2:211-33

Walsh, J.J., The biological consequences of interaction of the climatic, El Niño, and event scales
1978   of variability in the Eastern Tropical Pacific. Rapp.P.-V-Réun.CIEM, 173:182-92

_____, A carbon budget for overfishing off Peru. Nature,Lond., 290:300-304
1981

Walt, K.E.F., Ecology and resource management. New York, McGraw-Hill Book Company, 450 p.
1968

Ware, D.M., Growth, metabolism and optimal swimming speed of a pelagic fish. J.Fish.Res.Board
1975   Can., 32(1):33-41

_____, Bioenergetics of stock and recruitment. Can.J.Fish.Aquat.Sci., 37(6):1012/24
1980

Webb, P.W., Hydrodynamics and energetics of fish propulsion. Bull.Fish.Res.Board Can., (190):
1975   159 p.

Weihs, D., Optimal fish cruising speed. Nature, Lond., 245:48-50
1973

_____, Energetic significance of changes in swimming modes during growth of larval anchovy,
1980   Engraulis mordax. Fish.Bull.NOAA/NMFS, 77:597-604

Welcomme, R.L., Fisheries ecology of floodplain rivers. London, Longman, 317 p.
1979

Welcomme, R.L. and H.F. Henderson, Aspects of the management of inland waters for fisheries. FAO
1976   Fish.Tech.Pap., (161):40 p. Issued also in French and Spanish

Whittaker, R.H., Gradient analysis of vegetation. Biol.Rev.Camb.Philos.Soc., 49:207-64
1967

Wildish, D.J. and R.L. Phillips, An identification strategy for benthos collected to assess marine
1974   and estuarine pollution. Tech.Rep.Fish.Res.Board.Can., (450):31 p.

Williams, F., Food of longline-caught yellowfin tuna from East Africa Waters. E.Afr.Agric.For.J.,
1966   31(4):375-82

Winberg, G.G., Rate of metabolism and food requirements of fishes. Transl.Ser.Fish.Res.Board
1960   Can., (194):202 p.

Windell, J.T., Digestion and the daily ration of fishes. In Ecology of freshwater fish produc-
1978   tion, edited by S.D. Gerking. New York, John Wiley and Sons, pp. 159-83

Yokota, T., et al., Studies on the feeding habit of fishes. Rep.Nankai Reg.Fish.Res.Lab.,(14):
1961   234 p. (in Japanese, English summary)

Young, D.K., M.A. Buzas and M.W. Young, Species densities of macrobenthos associated with sea-
1976   grass: a field experimental study of predation. J.Mar.Res., 34(4):577-92

Zaret, T.M., Predation and freshwater communities. New Haven, Connecticut, Yale University Press,
1980   187 p.

Anon., Why bother with marine conservation? Conserv.Indon., 5(3)
1981

M-43
ISBN 92-5-102510-X

Tip. INTERSTAMPA r.l. - Via Barbana, 33 - Tel. 06/540.33.49